Wireless Networks

Series Editor

Xuemin Sherman Shen, University of Waterloo, Waterloo, ON, Canada

The purpose of Springer's Wireless Networks book series is to establish the state of the art and set the course for future research and development in wireless communication networks. The scope of this series includes not only all aspects of wireless networks (including cellular networks, WiFi, sensor networks, and vehicular networks), but related areas such as cloud computing and big data. The series serves as a central source of references for wireless networks research and development. It aims to publish thorough and cohesive overviews on specific topics in wireless networks, as well as works that are larger in scope than survey articles and that contain more detailed background information. The series also provides coverage of advanced and timely topics worthy of monographs, contributed volumes, textbooks and handbooks.

Xiaoming Tao • Yiping Duan • Zhijin Qin •
Danlan Huang • Liting Wang

Wireless Multimedia Computational Communications

Springer

Xiaoming Tao
Department of Electrical Engineering
Tsinghua University
Beijing, China

Zhijin Qin
Department of Electrical Engineering
Tsinghua University
Beijing, China

Liting Wang
Department of Electrical Engineering
Tsinghua University
Beijing, China

Yiping Duan
Department of Electrical Engineering
Tsinghua University
Beijing, China

Danlan Huang
School of Information and Communication Engineering
Beijing University of Posts and Telecommunications
Beijing, China

ISSN 2366-1186 ISSN 2366-1445 (electronic)
Wireless Networks
ISBN 978-3-031-64154-1 ISBN 978-3-031-64155-8 (eBook)
https://doi.org/10.1007/978-3-031-64155-8

This Springer imprint is published by the registered company Springer Nature Switzerland AG
The registered company address is: Gewerbestrasse 11, 6330 Cham, Switzerland

Foreword

According to the Cisco Visual Networking Index published in 2019, by 2022, video will account for 82% of all IP traffic. Looking back, the global IP video traffic has grown threefold from 2016 to 2021, with a CAGR of 26 percent. Internet video traffic has grown fourfold from 2016 to 2021, a CAGR of 31 percent. It is expected that such growth will continue at the same or even higher rate in the next few years.

Such a drastic growth is particularly noticeable after the pandemic in 2020 when a large number of activities needed to shift from offline mode to online mode. This unexpected shift has greatly pushed the emergence of video chatting, video games, and video information flow to meet both working and leisure needs. Both visual networking developers and ordinary visual data users can feel strongly that video networking has become a new and important way of interaction for both professional activities and daily life routines.

At the device and connection fronts, the rise of video-empowered new capabilities has been greatly accelerated and, according to Cisco predictions, we will soon see at least half of these devices and connections have been equipped with video capabilities for various visual data-driven services. It is expected that seamless visual data networking shall enable users and developers to adapt to the rapidly changing world more swiftly, enhancing the importance of real-time video applications and remote presentations.

As visual data are commonly augmented with audio, text, graphics, and other types of media data, video data networking has already been broadened to the emerging multimedia data networking to handle multiple media data types simultaneously, rather than the conventional single type of video data for both algorithm design and user applications.

On the networking front, among all types of connection means, wireless networking is the most challenging type of connection for mobile devices because wireless channels are always changing caused by their unfriendly and less predictable networking environments. When massive amounts of visual data generated by various contemporary applications are required for transporting over wireless networks, it is still extremely demanding to ensure robust streaming over wireless links and connections even though we are in the era of 5G. This is particularly true when the objective of visual data streaming is indeed to convey the semantics embedded in massive visual data.

All these advances and requirements of multimedia networking suggest that contemporary visual networking would be fundamentally different from these conventional approaches developed since Shannon first published his revolutionary paper on the mathematical theory of communication more than 70 years ago. Shannon's paper laid the foundation for several past generations of digital communication and networking in which we have been focusing on voice and data transmission. In the past, we assumed that source data could be properly converted into discrete symbols for coding and transmission. With rich semantics embedded in the visual data, conventional source coding will need to be completely transformed to represent the rich semantics embedded in the visual data. Therefore, new computational tools for extracting and representing visual semantics from source data will need to be developed to ensure seamless and effective transmission of massive visual data from sources to destinations via wireless networking. We are expecting a new paradigm of multimedia networking in which computational resources at both the source and the destination of the networking continuum can be exploited to enhance the overall performance of multimedia data networking.

With the technical challenges and new demands in visual data networking presented above, a new book that illustrates the new fundamental principles in wireless multimedia will be an invaluable reference for researchers and application developers who are working on this emerging technical front. With such needs in mind, I am very happy to introduce this new book, *Wireless Multimedia Computational Communications*, authored by members of the research group from Tsinghua University. This group has been conducting numerous pioneering researches in mobile networking, especially in wireless multimedia, and I also have a long history of collaboration with this group.

This book presents some contemporary designs and applications with a focus on multimedia content delivery over wireless links based on computational communication. It begins with a comprehensive overview of the new architecture and fundamental principles of multimedia computational communications. The objective of computational communication is to enable new capabilities in wireless multimedia through a balanced allocation of resources for computational tasks and communication tasks. Such a computational communication approach is expected to open up new research avenues for wireless multimedia networking.

The authors of this book have been able to present several unique perspectives in the emerging field of wireless multimedia and have organized their book in a well-structured fashion to help the readers get familiar with these technical challenges and potential solutions quickly. In particular, the book presents the key technologies that have been successfully developed by the research team. The authors explain clearly the key techniques that underpin these processes, highlighting the ability to process vast amounts of data, create seamless connections, and design algorithms that can learn and evolve under adversary networking environments. Furthermore, the authors also introduce user experience based on electroencephalography as an effective evaluation criterion for visual data coding and communication.

This book also discusses how to design a more intelligent and adaptive wireless network by utilizing a prior knowledge base that can be derived from previously acquired information. These networking-oriented investigations are based on neural networks designed for structural coding in various scenarios with convincing results obtained based on several contemporary datasets. The book also presents end-to-end semantic communication frameworks for multimodal and multitasking applications. It addresses optimization issues for the quality of experience, focusing on rate-semantic distortion and rebuffering through pricing and electroencephalography modeling. One highlight of this book is the presentation of intelligent coordination and computing across the cloud, edge, and terminal for supporting proposed computational communication frameworks. This book concludes with a discussion of future research directions, including brain-inspired multimedia computational communications and semantic communications, offering valuable insights into the great potential of wireless multimedia.

I would like to emphasize that this book is written concisely and clearly. It is an excellent reference for broad audience including graduate students, faculty researchers, and practice engineers.

Chair Professor of Visual Computing, The Hong Kong Polytechnic University
December, 2023

Changwen Chen

Preface

In this book, we will discuss the evolving designs and applications of multimedia content delivery, focusing on computing-based methods. Our aim is to offer readers an in-depth understanding of how advanced computational methods can help multimedia content delivery. We start with an overview of the principles of multimedia computational communications and then detail the emerging computational techniques that enable the processing of large data volumes and the design of adaptive algorithms. We introduce user experience evaluation using electroencephalography and discuss the application of neural networks in various scenarios, supported by real-world datasets. Then, the semantic knowledge base with the next generation multiple access is proposed to achieve semantic fidelity in reconstructing high-resolution, stable, and smooth video data. We also present an end-to-end semantic communication framework for multitasking applications and discuss the optimization of user experience by addressing rate-semantic distortion and rebuffering issues. A further section delves into intelligent coordination across cloud, edge, and end for supporting these frameworks. In the final part of the book, we look into the future, focusing on exciting developments such as brain-inspired multimedia computational communications and semantic communications. We believe this book will provide an insightful understanding of multimedia content delivery through advanced computational methods.

Beijing, China
August, 2023

Xiaoming Tao

Acknowledgments

I would like to express my sincere gratitude to my current and former PhD students, my collaborators, and all the colleagues who contributed to the work and projects that lead to this book.

I would also like to particularly thank our editor as well as all the editorial staffs from Springer to produce this book.

Finally, I would like to thank my family. I cannot make it without their great support.

This work was supported by the National Natural Science Foundation of China (Nos. NSFC 61925105, 62227801, 62293484, 62322109, and 62171257), the State Key Laboratory of Space Network and Communications, Beijing National Research Center for Information Science and Technology, New Cornerstone Science Foundation through the XPLORER PRIZE, and the Tsinghua University (Department of Electronic Engineering)—Nantong Research Institute for Advanced Communication Technologies, Joint Research Center for Space, Air, Ground and Sea, and Cooperative Communication Network Technology.

Beijing, China
August, 2023

Xiaoming Tao

Contents

Acronyms

APP	Application
AWGN	Additive white Gaussian noise
BERT	Bidirectional Encoder Representations from the Transformers
BLEU	Bilingual evaluation understudy
BS	Base station
BSI	Buffer state information
CE	Cross entropy
CNN	Convolutional neural network
CSI	Channel state information
DCT	Discrete cosine transform
DDPG	Deep deterministic policy gradient algorithm
DGL	Deep Graph Library
DL	Deep learning
DNN	Deep neural network
DSG	Deep sketch graph
EEG	Electroencephalography
ERP	Event-related potential
FC	Fully connected
FFT	Fast Fourier Transform
FIR	Finite impulse response
FSS	Feature subset selection
GNN	Graph neural network
IaaS	Infrastructure as a Service
ICA	Independent component analysis
IoT	Internet of Things
IQA	Image quality assessment
ISC	Image semantic communications
JSC	Joint source-channel
JSCC	Joint source-channel coding
KID	Kernal-Inception distance
KPI	Key performance indicators

KQI	Key quality indicators
LSTM	Long short-term memory
MAE	Mean absolute error
mAP	mean average precision
MDP	Markov decision process
MIMO	Multiple-input multiple-output
MLPS	Multi-layer perceptrons
MOS	Mean opinion score
MRC	Maximal ratio combining
MSE	Mean squared error
NGMA	Next-generation multiple access
NMS	Non-maximal suppression
OAR	Object-attribute-relation
PaaS	Platform as a Service
PSNR	Peak signal-to-noise ratio
QoE	Quality of Experience
QoS	Quality of Service
RD	Rate-distortion
RDA	Rate decision algorithm
RL	Reinforcement learning
RNN	Recurrent Neural Network
SaaS	Software as a Service
SINR	Signal-to-interference-plus-noise ratio
SNR	Signal-to-noise ratio
S-R	Semantic transmission rate
S-SE	Semantic spectral efficiency
SSIM	Structural similarity
SVM	Support vector model
TSC	Text semantic communications
VQA	Visual Question Answering
VQEG	Video quality expert group
b	The index of cells in cellular networks
b_m^l	The bias of the m-th node in the l-th layer
β	The learning rate
$cov(\cdot,\cdot)$	The covariance
$d_{()}$	The distance
$\mathcal{D}$	The dataset
$F_{()}$	The features
$f(\cdot)$	The functions
γ	The signal-to-noise ratio
g	The large-scale channel gain
H_{image}	The height of an image
$\mathbf{H}$	The channel matrix
$\Delta\mathbf{H}$	Channel estimation error
$\widehat{\mathbf{H}}$	The estimated CSI

$\mathbf{h}$	The small-scale fading coefficient
I	The image
I_{sem}	The semantic information
i, j	The index of users
$\langle \cdot \rangle$	The inner product
k	The index of channels
k_h	The index of the heatmap
l	The index of layers
L	The loss of the neural network or gradient
$\|\cdot\|_0$	The l_0 norm
m, n	The index of nodes
μ	Mean value
$N_{()}$	The quantity
$\mathbf{N}$	The Additive white Gaussian noise
o	The index of words
ω	The weights
$P(\cdot)$	The probability
p	The transmit power
φ	The bit rate
φ_{sem}	The semantic rate
Q	The quartile
$\mathcal{Q}$	The modalities
ρ	The Pearson correlation coefficient
ρ_s	The Spearman correlation coefficient
S	The scores
s	The index of sentences
$\boldsymbol{s}$	The data source
σ	The variance
t	The index of time and steps
u	The index of sentences
W_{image}	The width of an image
X_f	The feature vector
$\boldsymbol{x}$	The transmitted signal
$\hat{\boldsymbol{x}}$	The recovered signal
Y_s	The score vector
$\mathbf{Y}$	The received signal
(x_s, y_s)	The coordinate of s-th sampled point
$\Delta x_s, \Delta y_s$	The amount of coordinates change
$z, \hat{z}$	The real and estimated semantic information

Chapter 1
Wireless Multimedia Computational Communications

1.1 Introduction

Multimedia, a term that has become integral to our understanding of the digital age, represents the convergence of text, audio, images, video, and animation into a single platform. It permeates our daily lives and allows us to control how we consume content, as well as navigate through the information according to our preferences. The volume of data is increasing exponentially, spurred by the proliferation of connected devices, the rapid expansion of the Internet of Things (IoT), and the surge in multimedia content production and consumption. Traditional data transmission and processing methods are being pushed to their limits, signaling the need for a new paradigm.

This emerging paradigm revolves around leveraging computational communications to manage, process, and optimize the deluge of data in our wireless multimedia systems. By integrating advanced computing techniques—such as machine learning, data analytics, and cloud computing into our wireless networks, we can make networks smarter, faster, and more efficient. This new era of wireless multimedia, underpinned by computational capabilities, holds the potential to revolutionize our data handling practices, reshaping the future of communication, entertainment, education, and beyond.

Alongside the developments in artificial intelligence (AI), this book spans the framework of wireless multimedia communication powered by computing. According to Shannon and Weaver's theory, the first level of communication is syntactic communication, where the communication system serves as a pipeline for accurately transmitting information symbols. As we look back at the development of wireless multimedia design, the goal of transmission was to accurately recover the signal transmitted by content providers, through scheduling, source coding, channel coding, and resource allocation. However, as multimedia has become more complex and user-centric, we need to reevaluate the technical efficiency of multimedia systems. This book presents each link in the communication system with case

X. Tao et al., *Wireless Multimedia Computational Communications*,
Wireless Networks, https://doi.org/10.1007/978-3-031-64155-8_1

studies and matching algorithms to demonstrate how emerging AI technologies can benefit wireless computing multimedia systems.

1.2 Architecture of Multimedia Computational Communications

1.2.1 Evaluation Criterion: QoS → QoE

Quality of service (QoS), including metrics like peak signal-to-noise ratio (PSNR) and packet loss rate, is an objective evaluation criterion premised on microscopic precision. Its optimization is increasingly difficult and often constrains the performance of transmission and network optimization. Quality of experience (QoE), on the other hand, targets macro satisfaction factors such as attention, clarity, and smoothness and can relax the requirement of precision, satisfying users' subjective feelings in an approximate way. The study of QoE involves the intersection of two fields: informatics and visual psychology. It is an emerging frontier direction that faces challenges such as how to measure and optimize QoE within system design.

1.2.2 Prior Knowledge Base and Structural Coding: Pixel → Spatial-Temporal Structure

With the growth of multimedia services, the assessment of multimedia communication quality is gaining importance. The traditional assessment metrics for communication systems are based on the QoS. However, due to the increase in multimedia data and the diversity of services, the QoS metrics, including packet loss, bit rates, PSNR, etc., are becoming less suitable for matching different multimedia applications and often inconsistent with real human experience. QoE emphasizes the user's perceived satisfaction, which is user- and perception-centered, focusing on the human's actual experience. Therefore, QoE needs to be introduced into modern communication systems. With the progress of psychophysiology and neuroscience, electroencephalography (EEG) is gaining more attention in the fields of communications and signal processing. EEG signals are recorded through a noninvasive electrophysiological device that captures the electrical activities of the brain by deploying electrodes on the scalp. The EEG method offers high temporal resolution and typically measures implicit rather than explicit responses to physical stimuli. Thus, by analyzing the acquired EEG signals, researchers can extract the perceptual information of the subject from the complex neural activity. Given these characteristics of EEG, we are able to capture bottom-level information and find many important regions and reaction mechanisms related to quality assessment by mining and learning from human EEG responses. Therefore, it is possible to harness

EEG signals and the associated mechanisms to construct new assessment metrics and models to more accurately measure humans' real perceptions.

1.2.3 Semantic Information Transmission: Rate-Distortion → Rate-Semantic-Distortion

In traditional communications, the rate-distortion theory gives an analytical expression for how much compression can be achieved using lossy compression methods. The rate-distortion theory can also connect the distortion to the channel capacity [2], qualifying the required bits per symbol that can be transmitted over the channel without exceeding an expected distortion. The distortion [1] is measured by the distance between the original signal and the reconstructed signal, such as the mean squared error (MSE), PSNR, and other perceptual metrics based on human perception.

In semantic information transmission, according to different transmission goals, the receivers can not only reconstruct the signal but also restore the semantic information of the signal, such as the label and scene of the image [3]. The semantic information of the signal may not always linearly correlate with the reconstructed signal. Therefore, the rate-distortion theory cannot accurately qualify and measure the required bits per symbol of semantic information based on conventional distortion. In order to provide an analytical expression of how much compression can be achieved using semantic compression methods, it is necessary to extend the rate-distortion theory to the rate-semantic-distortion theory. This extraction enables the rate-semantic-distortion theory to qualify the required bits per symbol for the expected conventional distortion and semantic distortion. The semantic distortion can be measured by the distance between the semantic information contained in the original signal and the reconstructed semantic information, which depends on factors such as classification accuracy and retrieval recall.

The rate-semantic-distortion theory helps evaluate the rate and distortion performance of semantic communication systems. Specifically, when connecting the rate-semantic-distortion theory to physical channels, it can measure the required bits per symbol as well as the semantic distortion without exceeding the channel capacity. This indicates how many semantic bits can be transmitted over the air regarding the semantic distortion. For example, given the channel capacity C, we can compute the required bits per symbol $R(D)$ corresponding to semantic distortion D, which satisfies the condition $R(D) < C$. This can be formulated as a rate-semantic-distortion optimization problem. More information can be found in Chap. 6.

Table 2.4 Standard of data cleaning for four user scores. There is a notable correlation between the data and the scores, and these data are relatively manageable

Scores	Parameters			
Visual quality	Video play total time	User buffer time	Video peak rate	Video average rate
Loading quality	User buffer time	Buffer delay	Video play total time	Bit rate
Stalling quality	User buffer time	Video play total time	Bit rate	Buffer delay
Overall quality	Bit rate	Video size	Video average rate	Buffer delay

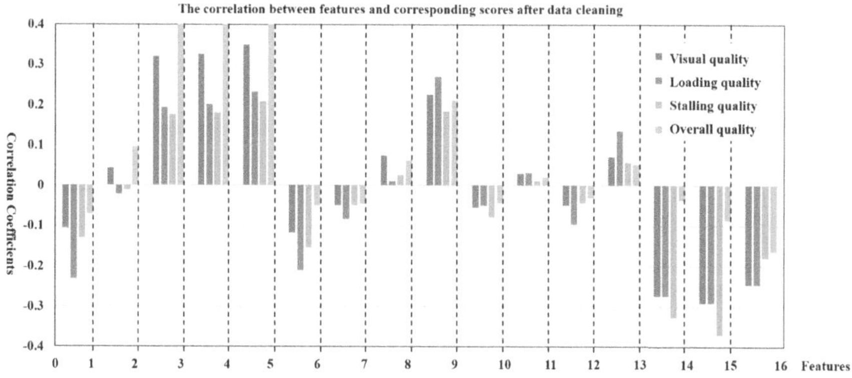

Fig. 2.10 Correlation coefficients between the features and the corresponding scores after data cleaning. The blue, red, green, and yellow bars represent the correlations between the features and the visual quality, loading quality, stalling quality, and overall quality, respectively. For each feature, we compute four correlations between the feature and its corresponding four scores. Features are labeled as 1-16: 1. user buffer time, 2. buffer delay, 3. bit rate, 4. video size, 5. video average rate, 6. video play total time, 7. screen resolution long, 8. video peak rate, 9. video all peak rate, 10. light intensity, 11. phone screen brightness, 12. ping average rate, 13. signal strength, 14. video stalling number, 15. video stalling total time, and 16. video stalling duration proportion. Moreover, a positive correlation indicates a positive relationship. The magnitude of the correlation coefficient reflects the strength of the correlation, whether positive or negative. The absolute value of the correlation coefficient does not exceed one

outliers from the selected database after FSS. The box plot method is used for data cleaning. To validate the performance of the data cleaning process, we compute the correlation between the scores and the cleaned data. The corresponding results are presented in Fig. 2.10. Figure 2.10 demonstrates the effectiveness of data cleaning. Furthermore, it reveals more pronounced correlations in the cleaned data, including obvious positive or negative correlations. This suggests the presence of numerous wild values in the original data that disrupted the data structure. The improved correlations validate the correctness and utility of the feature selection process. The cleaned data enhance the construction and understanding of the DNN model.

Moreover, it is important to note the difference between Figs. 2.9 and 2.10. Initially, we collected 80,000 user data points with 89 parameters. However, each

1.2.4 Cloud-Edge-End Intelligent Collaboration and Computing

In recent years, cloud intelligence has emerged as a potent enabler of intellectualization in daily life, as evidenced by technologies like facial payment. However, its disadvantages have become more pronounced, such as difficulty in achieving differentiation, an inefficient input-output ratio, and limited practical application effects. To address these challenges, there is a growing trend toward a better computing scheme: cloud-edge-end intelligence.

Cloud-edge-end collaboration is becoming increasingly popular due to its efficiency and flexibility, allowing it to adapt to different scenarios and improve the user experience. By placing computing at the front end and leveraging cloud-edge intelligence, it can perform accelerated computation based on the specific requirements of each scenario, facilitating faster data processing and reducing latency. Additionally, it can provide better privacy and security compared to traditional cloud computing solutions by keeping data and computing resources closer to the edge, thereby reducing the risk of data breaches or other security threats.

An important example is the active vision camera, which offloads computation to the edge, adapts to the scene for active perception and calculation, and processes key information locally on the camera side. With the support of intelligent sensing technology and efficient computing power, the active vision camera is capable of fulfilling more advanced requirements, such as rapid warning, real-time interaction, and real-time control, in addition to traditional shooting and recording functions. Cloud-edge-end intelligence is a promising new approach to computing that has the potential to revolutionize our interaction with technology and improve our lives in countless ways.

References

1. Blau, Y., Michaeli, T.: Rethinking lossy compression: The rate-distortion-perception tradeoff. In: Chaudhuri, K., Salakhutdinov, R. (eds.) Proceedings of the 36th International Conference on Machine Learning, ICML 2019, 9–15 June 2019, Long Beach, California, Proceedings of Machine Learning Research, vol. 97, pp. 675–685. PMLR (2019)
2. Davisson, L.D.: Rate distortion theory: a mathematical basis for data compression. IEEE Trans. Commun. **20**(6), 1202 (1972)
3. Qin, Z., Ye, H., Li, G.Y., Juang, B.F.: Deep learning in physical layer communications. IEEE Wirel. Commun. **26**(2), 93–99 (2019)

Chapter 2
QoE Evaluation Model Based on EEG

2.1 Data-Driven QoE Evaluation Model: From Network Parameters to QoE

Quality of experience (QoE) serves as a direct evaluation of users' experiences in mobile video transmission and is essential for network management, such as network optimization. In this chapter, we propose a deep learning–based QoE prediction approach using a large-scale QoE dataset for mobile video transmission. Specifically, we developed a mobile phone app for collecting user QoE data when users are viewing videos transmitted over the mobile Internet in a real-world environment. Subsequently, we construct a large-scale dataset by collecting over 8000 data points with four types of subjective scores and 89 network parameters. Each QoE metric is only related to some of the 89 network parameters. Therefore, we apply a feature selection method to identify the feature parameters relevant to user scores. Additionally, we employ the box plot method to clean the raw data by removing outliers. Finally, we develop a deep neural network (DNN) to learn the relationships between the network parameters and the subjective QoE scores. The proposed DNN can also be viewed as a data-driven objective QoE prediction approach for mobile video transmission, which can predict user QoE scores. Experimental results demonstrate that the proposed approach effectively eliminates most features that are irrelevant to QoE prediction and outperforms other state-of-the-art approaches in terms of QoE prediction performance.

2.1.1 Introduction

In recent years, there has been an explosive increase in the number of mobile terminal users [33], resulting in a significant surge in mobile data traffic. In the upcoming 5G era, mobile video transmission is expected to be a key service,

X. Tao et al., *Wireless Multimedia Computational Communications*,
Wireless Networks, https://doi.org/10.1007/978-3-031-64155-8_2

accounting for the majority of mobile data traffic. As a result, mobile video transmission has become an active research topic in wireless communications. It is particularly important to meet the quality expectations of terminal users in mobile video transmission. Therefore, it is crucial to assess the QoE for mobile video transmission [15, 25, 26, 35, 36, 53, 66].

The motivations behind our proposed approach are summarized as follows: First, in the upcoming 5G era, video services will be the fundamental service accounting for the majority of mobile data traffic. Consequently, more attention should be paid to improving user experience in future network management, which is the ultimate goal of mobile network services. Therefore, it is necessary to analyze QoE when using network services. Second, traditional QoE prediction methods often rely on predefined or empirical functions to represent the relationships between mobile video transmission parameters and subjective assessments of QoE. However, these relationships are inaccurate. Data-driven methods offer promising approaches to predict QoE more accurately. It is widely recognized that high-quality data is the foundation of data-driven methods. Therefore, collecting practical user data is essential. Moreover, the relationships between mobile video transmission parameters and subjective assessments are complex. Therefore, it is necessary to develop adaptive learning methods to capture these relationships. Based on these considerations, we have developed an app to collect data from a large number of users. Subsequently, a DNN is employed to learn the relationships between the network parameters and the subjective QoE scores.

In this chapter, we propose a deep learning–based QoE assessment approach using a large-scale QoE dataset for mobile video transmission. Specifically, we have developed a video collecting application (app) to gather user QoE data when users are viewing videos through the mobile Internet in a real-world environment, as opposed to a laboratory environment used in previous approaches. With the app, we have constructed a dataset consisting of QoE data from 80,000 samples. Each user's QoE data include subjective scores for four metrics: visual quality, loading, stalling, and overall quality. Additionally, we record 89 network parameters related to mobile video transmission. Each QoE metric is only associated with a subset of the 89 network parameters. To address this, we have employed a feature selection method [48, 49] to reduce the redundancy of the network parameters for QoE assessment. Furthermore, we have utilized the box plot method to clean the raw data by removing outliers. Finally, a DNN is developed to learn the relationships between the network parameters and the subjective QoE scores. The framework of the proposed approach is illustrated in Fig. 2.1, which encompasses data collection, data analysis, and model learning. The proposed DNN can also be considered a data-driven QoE prediction approach. The main contributions of this chapter are twofold:

1. We establish a large-scale dataset consisting of 80,000 QoE scores obtained from users viewing videos over the mobile Internet by developing a new mobile phone app.

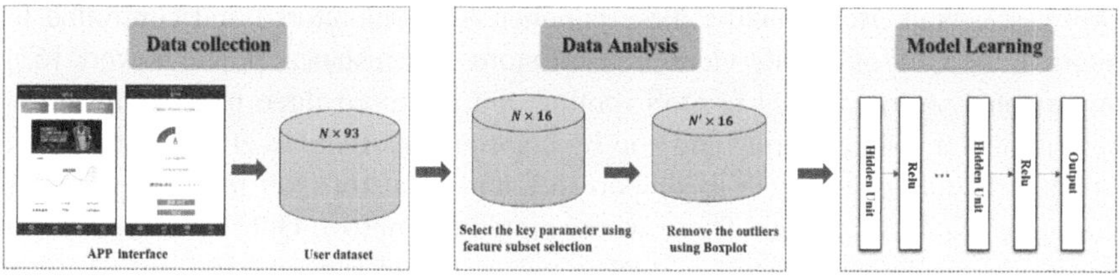

Fig. 2.1 Framework of the proposed approach. In data collection, the left picture shows the interfaces of our developed APP. The dataset size is $N \times 93$. In data analysis, we initially select 16 key network parameters, resulting in a dataset size of $N \times 16$. Subsequently, we remove the outliers, resulting in a dataset size of $N' \times 16$. In model learning, we design network structures and a loss function to train the QoE model

2. We develop a new DNN approach to learn the relationship between network parameters of mobile video transmission and subjective QoE scores, enabling the prediction of subjective QoE scores. This approach can be viewed as a data-driven QoE prediction approach.

This chapter is organized as follows: Sect. 2.1.2 reviews related work on QoE prediction. Section 2.1.3 describes how we construct a mobile video transmission user database and provides a thorough analysis of the database. In Sect. 2.1.4, we propose a new DNN approach to predict subjective QoE scores, including the design of the network architecture and the loss function. Section 2.1.5 presents the experimental results, and Section VI concludes this chapter.

2.1.2 *Related Work*

Several approaches have been proposed to predict the QoE of mobile video transmissions [5, 8, 28, 34, 43], mainly falling into four categories: QoS monitoring, subjective assessment, heuristic objective assessment, and data-driven objective quality assessment. Subjective assessment is considered the ultimate QoE assessment as it directly reflects users' opinions when viewing videos. However, it is time-consuming and impractical to implement. QoS monitoring and heuristic objective assessment have been extensively studied but have limitations in accurately capturing the relationship between mobile video transmission parameters and subjective QoE assessments. These parameters typically include key quality indicators (KQI) and network key performance indicators (KPI). The fourth category, data-driven objective quality assessment, has gained increasing research interest as it focuses on exploring the relationship between mobile video transmission and QoE from a data-driven perspective.

QoS monitoring [1, 7, 62] involves using video transmission network parameters to assess end users' QoS directly. These parameters are selected based on prior knowledge of video transmission and are used to establish predefined functions for

QoS assessment. Additionally, these parameters are monitored and controlled to improve the QoS of mobile video transmissions. For instance, previous work [62] investigated various issues in QoS routing and proposed three path computation algorithms for source routing and hop-by-hop routing. Andrews et al. [1] discussed simple scheduling strategies to support different forms of QoS for multiple users sharing a wireless channel. These works aimed to improve QoE by appropriately selecting parameters relevant to QoE, such as video compression optimization and network bandwidth allocation. Although monitoring and controlling QoS are important aspects of mobile video transmission, they do not solely determine end users' QoE. QoE is also influenced by subjective factors related to video viewing, making subjective assessment necessary for evaluating the QoE of mobile video transmission.

Subjective assessment [6, 63] refers to directly asking volunteers to assess QoE in viewing videos. Volunteers are required to watch a series of video sequences and rate their corresponding scores. By analyzing the rated scores of different video sequences, researchers can investigate the relationship between mobile video transmission network parameters and subjective scores. For example, a study [63] reviewed several subjective image and video quality assessment approaches, including a new paradigm for quality assessment based on structural distortion measures, which offer advantages over traditional perceptual error sensitivity-based approaches. However, subjective assessment is typically conducted in a laboratory environment, which may not accurately represent real-life scenarios of video viewing. Moreover, subjective assessment is time-consuming and cannot be performed in real time. Therefore, objective assessment methods are required for more practical implementation.

In contrast to subjective assessment, numerous objective assessment approaches [37, 64, 68] have been proposed. Early objective QoE assessment methods were heuristic in nature. In this category, machine-based methods automatically measure the QoE of mobile video transmission with the aim of optimizing the correlation with subjective assessment scores. For instance, a content-adaptive packet-layer model was designed in a study [64], which incorporated estimates of temporal complexity and frame type, making it suitable for quality assessment of individual services. Another work [37] introduced a novel no-reference video quality assessment algorithm in the compressed domain. Unlike conventional video quality assessment methods that assess video quality frame by frame, this method processes multiple video frames together to obtain an objective quality score. Generally, certain parameters related to mobile video transmission are selected, and mathematical functions, such as composite functions, are formulated to assess the QoE of received videos. However, these mathematical functions are still in their early stages, as they fail to accurately capture the subjective factors of QoE. In contrast, large-scale user data are powerful for learning the relationship between video transmission metrics and QoE. Therefore, data-driven objective quality assessment shows promise for evaluating the QoE of mobile video transmission.

Recently, several data-driven objective quality assessment approaches have been proposed [5, 9, 47]. These approaches learn the correlation between video transmission parameters and QoE from datasets that include subjective QoE scores of video sequences and their corresponding parameters. For example, Gadaleta et al. [17] presented D-DASH, a framework that combines deep learning and reinforcement learning techniques to optimize the QoE of dynamic adaptive streaming over HTTP (DASH). In another study [21], novel QoE-driven download resource allocation algorithms were investigated for green solutions in content-centric IoT to enhance QoE satisfaction. Additionally, a high-accuracy allocation algorithm was proposed to manage IoT from an experiential standpoint and improve the close-to-green solution using deep Q-learning. To achieve optimal performance across various network conditions and QoE objectives, the work [42] proposed Pensieve, a system that generates adaptive bit rate (ABR) algorithm using reinforcement learning (RL). Pensieve trains a neural network model to select bit rates for future video chunks based on observations collected by client video players. More recently, deep learning has demonstrated remarkable capabilities in massive data analysis and processing [23, 45, 61]. Similarly, data-driven objective quality assessment can benefit from the recent success of deep learning. However, datasets for QoE assessment of mobile video transmission currently lack sufficient data for deep learning purposes.

2.1.3 Database Establishment and Analysis

One of the challenges in data-driven QoE assessment, especially when employing deep learning, is the scarcity of training data. Therefore, it is crucial to establish a large-scale QoE database to facilitate research in QoE assessment. This section primarily focuses on the establishment and analysis of our extensive QoE database.

2.1.3.1 Data Collection

APP To collect data from a large number of users, we have developed a mobile phone app called CU-vMOS, supported by China Unicom. We have deployed CU-vMOS in two provinces in China through collaboration with China Unicom. Figure 2.2 illustrates the interfaces of our app. Figure 2.2a displays the start interface, Fig. 2.2b shows the interface when the user watches a video, and Fig. 2.2c represents the user evaluation interface. Additionally, we have constructed a video database consisting of 240 videos, including movie clips, cartoons, and sports videos. Figure 2.3 displays examples of six videos from our video database. The data collection process can be divided into three steps: Step 1: user downloads the app from the app center; step 2: user opens the app and can access the video list. Each user will rate four scores, including visual quality, loading, stalling, and overall quality; and step 3: in the background of the app, we record 89 feature parameters and 4 user scores for each user. It is important to note that the 89 feature parameters are predefined. Initially, we aim to collect as many parameters as possible.

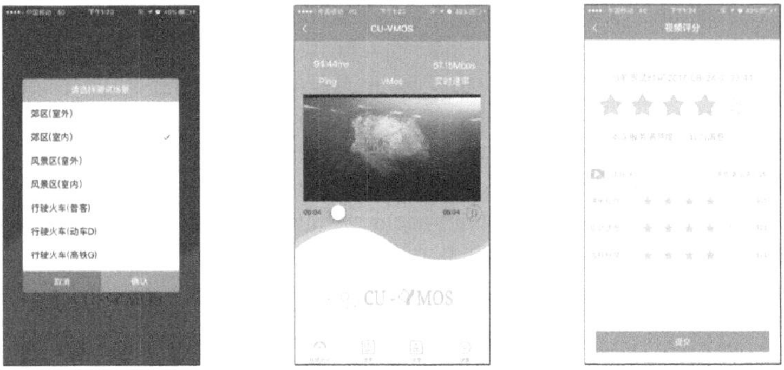

(a) (b) (c)

Fig. 2.2 Some interfaces of our APP. (**a**) represents the interface for scenario selection, (**b**) illustrates the interface when users watch videos, and (**c**) shows the interface for subjective scoring

Fig. 2.3 Examples from our video database

Database We invite volunteers to use CU-vMOS to watch videos and rate the QoE of the videos based on four metrics: visual quality, loading, stalling, and overall quality. These metrics directly reflect the user experience while watching videos. Specifically, visual quality measures the users' experience with video clarity, where HD and ultra HD videos are more likely to receive higher scores. The loading metric captures the users' experience with video start time delay, with longer startup times resulting in lower loading scores. The stalling metric refers to the users' experience with video buffering, as stalling occurs when the video needs buffering during playback. Lastly, the overall quality metric represents the users' satisfaction with the video quality. Each metric has five grades: 1, 2, 3, 4, and 5, which correspond to "very annoying", "annoying", "slightly annoying", "distortion perceptible, but not annoying", and "distortion imperceptible", respectively. Table 2.1 presents

Table 2.1 Definitions of the five grades

Scores	1	2	3	4	5
Meaning	Very annoying	Annoying	Slightly annoying	Perceptible, but not annoying	Imperceptible

the corresponding expressions. We record 89 network parameters, as shown in Table 2.2, which mainly include video coding format, video bit rate, and buffering delay, among others. During data collection from users in two Chinese provinces, we also obtain the corresponding network parameters through mobile terminals. Ultimately, we collect 80,000 QoE data points from users, where each data point consists of 89 feature parameters and 4 scores. The feature set is $89 \times N_{\text{data}}$, where N_{data} is the total number of data points. The total score set is $4 \times N_{\text{data}}$. Our objective is to identify the relationships between the feature set and the score set within this extensive user data. Table 2.3 provides detailed information about the dataset. The data distribution of the dataset is also illustrated in Fig. 2.4.

Currently, several databases exist for QoE assessment. For instance, the video quality expert group (VQEG) proposed a subjective test involving 287 viewers in [19], while the LIVE database [47] involved 31 viewers. However, these databases were constructed in a laboratory environment. In contrast, our databases are collected from real users in collaboration with China Unicom. Moreover, to the best of our knowledge, our database contains the most comprehensive set of network parameters.

2.1.3.2 Feature Selection

Within the collected network parameter database, there are redundant or irrelevant parameters with respect to QoE. In the data-driven analysis method, we consider network parameters as features in our approach. Therefore, we need to mine our dataset to identify the features relevant to QoE to accurately assess objective QoE. Several feature subset selection (FSS) methods [16, 48, 49] can be employed to reduce the dimensionality of QoE-related features. In our paper, we adopt the method described in [16], which utilizes bidirectional search to remove unrelated features. FSS selects a subset of features as the basis for training during model construction, aiming to optimize objective QoE assessment.

The implementation of feature selection generally involves the following components: generation process, evaluation function, stopping criteria, and verification process. The entire process is depicted in Fig. 2.5. Specifically, the original feature set serves as the input. For the generation step, we employ the bidirectional search method to generate the feature subset [22, 39]. Bidirectional search is a graph search algorithm that identifies the most related feature from an initial feature set to a goal subset within a directed graph. It runs simultaneous searches, one forward from the initial state, and one backward from the goal, halting when the two meet in

Table 2.2 89 network parameters and 4 scores

Serial number	DA_ID	Screen_resolution_long	Screen_resolution_width	Video_clarity
Video_coding_format	Video_bitrate	Buffering_delay	Video_buffering_peak_rate	Video_rebuff_num
Video_rebuff_total_time	Rebuff_duration_proportion	Rebuff_time_duration_one	Rebuff_time_point_one	Rebuff_time_duration_two
Rebuff_time_point_two	Rebuff_time_duration_three	Rebuff_time_point_three	Rebuff_time_duration_four	Rebuff_time_point_four
Rebuff_time_duration_five	Rebuff_time_point_five	Rebuff_time_duration_six	Rebuff_time_point_six	Rebuff_time_duration_seven
Rebuff_time_point_seven	Rebuff_time_duration_eight	Rebuff_time_point_eight	Rebuff_time_duration_nine	Rebuff_time_point_nine
Rebuff_time_duration_ten	Rebuff_time_point_ten	Video_total_time	Video_play_total_time	Video_all_peak_rate
Video_average_rate	Light_intensity	Phone_screen_brightness	Phone_place_state	Environmental_noise
Network_type	Phone_electric_start	Phone_electric_end	Signal_strength	Phone_version
Operating_system	Gps_longitude_one	Gps_latitude_one	Gps_longitude_two	Gps_latidude_two
Gps_longitude_three	Gps_latidude_three	Gps_longitude_four	Gps_latidude_four	Gps_longitude_five
Gps_latidude_five	Accelerometer_data	Instan_download_speed	User_score	Savedate
User_tel	Loading_score	Stalling_score	Vmos_score	CID
Lac	Userscene	Video_server_ip	Video_serve_location	Country
Province	City	Address	Ue_internal_ip	Elight
Estate	PLMN	IMSI	UDID	MCC
MNC	Cell_signal_strength	SINR	Ping_avg_rtt	RES1
CARRIER	LCID	Enodeg_id	Play_way	Starttime
Videosize	Videoname	Userbuffertime		

Table 2.3 Details of the dataset

Name	Total data number	Year	User score grade
QoE_dataset	80,000	2018	1, 2, 3, 4, 5
Video number	Video playback time	Number of feature parameters	Video resolution
240	30s	89	240p, 360p, 720p, 1080p

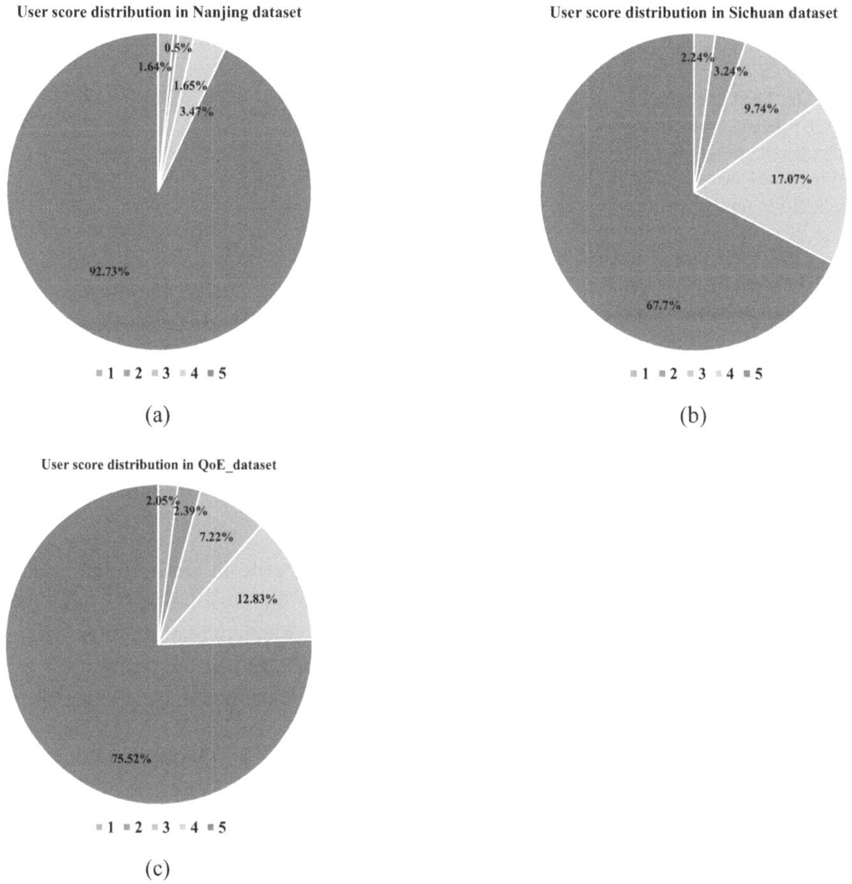

Fig. 2.4 User score distribution of the dataset. (**a**) shows the user score distribution in the Najing dataset, (**b**) depicts the user score distribution in the Sichuan dataset, and (**c**) represents the user score distribution of the entire dataset

the middle. Figure 2.6 illustrates the bidirectional search process, where the blue represents the complete feature set, the black circle represents the range of the unidirectional search, and the two red circles represent the range of the bidirectional search. It can be observed that the bidirectional search has a smaller search range compared to the unidirectional search, reducing the algorithm's complexity and time. The evaluation function serves as a criterion for evaluating the quality of

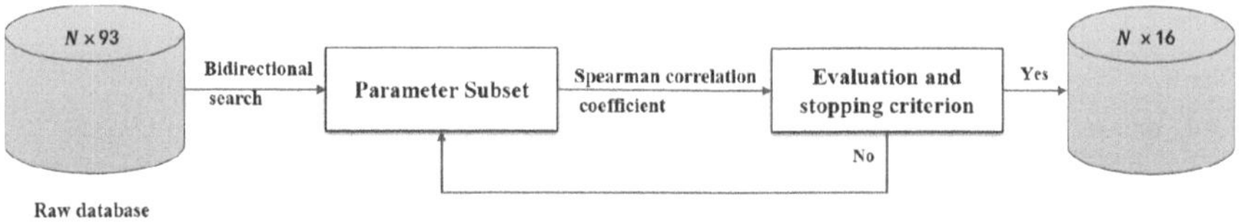

Fig. 2.5 Steps of the feature subset selection process. The FSS method is used to find the key parameters for QoE assessment from the raw database. N represents the total number of data points

Fig. 2.6 The bidirectional search process. The red part indicates the range of the bidirectional search, while the black circle represents the range of the unidirectional search. It is evident that the area of the red part is smaller than that of the black circle

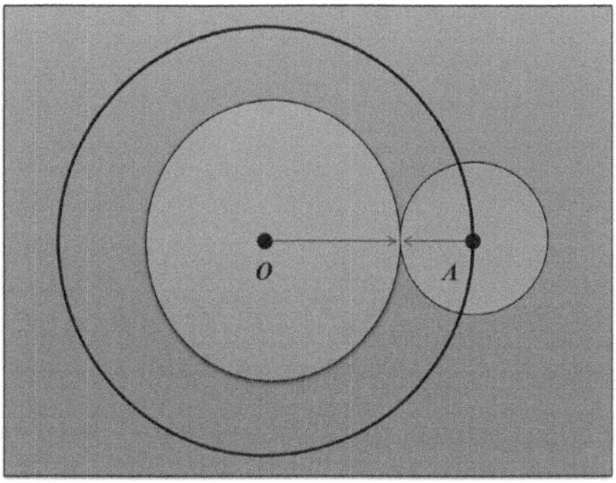

the generated feature subset. Two common evaluation criteria are correlation and distance. In the case of correlation, it is simple to use the correlation between the features and the scores to assess the quality of the selected feature subset. The characteristics of the higher output correlation are included in the feature subset. In contract, the characteristics of low correlation are eliminated. A good feature subset includes features that are strongly correlated with the scores. The Pearson correlation coefficient [52, 70], denoted as $\rho_{X,Y}$, is used as the evaluation function:

$$\begin{aligned}\rho_{X_f,Y_s} &= \frac{cov(X_f, Y_s)}{\sigma_{X_f}\sigma_{Y_s}} = \frac{E((X_f - \mu_{X_f})(Y_s - \mu_{Y_s}))}{\sigma_{X_f}\sigma_{Y_s}} \\ &= \frac{E(X_f Y_s) - E(X_f)E(Y_s)}{\sqrt{E(X_f^2) - E^2(X_f)}\sqrt{E(Y_s^2) - E^2(Y_s)}}\end{aligned} \tag{2.1}$$

where X_f is the feature vector and Y_s is the score vector. μ_{X_f} and μ_{Y_s} are the means of the vector X_f and Y_s. $E[\cdot]$ represents expectation and $cov(\cdot)$ represents the covariance of the elements. For simplicity, (2.1) can be written as:

$$\rho_{X_f,Y_s} = \frac{\sum X_f Y_s - \frac{\sum X_f \sum Y_s}{N_{\text{data}}}}{\sqrt{\left(\sum X_f^2 - \frac{(\sum X_f)^2}{N_{\text{data}}}\right)\left(\sum Y_s^2 - \frac{(\sum Y_s)^2}{N_{\text{data}}}\right)}} \tag{2.2}$$

Based on the Pearson correlation coefficient, the Spearman correlation coefficient [12] is defined as:

$$\rho_s = 1 - \frac{\sum_{i=1}^{N_{\text{data}}} d_i^2}{N_{\text{data}}(N_{\text{data}}^2 - 1)} \tag{2.3}$$

In the equation above, d_{rank} represents the difference in ranks when sorting two variables, X_f and Y_s. The rank is determined by the position of X_f and Y_s after sorting, and the difference between the two ranks is denoted as d_{rank}. N_{data} represents the number of data points.

The stopping criterion is closely related to the evaluation function and is generally determined based on the chosen evaluation function. In this chapter, we set a threshold to stop the feature selection process. Feature selection is an iterative process that continues until all features have been traversed. Through FSS, we have identified 16 critical network parameters for QoE prediction.

2.1.3.3 Data Cleaning

During the process of collecting user data, the users provide scores based on their subjective feelings, making outliers inevitable. Therefore, data cleaning or denoising is a crucial and indispensable step in data analysis. The outcome of data cleaning directly impacts the subsequent analysis and modeling processes. Effective data cleaning contributes to the construction of accurate models and predictions.

To address this issue, we employ the box plot [54] to remove outliers from the original data. The box plot, also known as the box-whisker plot, utilizes the minimum value, first quartile ($Q1$), median, third quartile ($Q3$), and maximum value of a set of data to depict the central position and spread range of the data distribution. $Q3 - Q1$ is defined as the interquartile range (IQR). There are two thresholds, $Q3 + 1.5IQR$ and $Q1 - 1.5IQR$, referred to as the outlier truncation points or internal limits. Additionally, there are two thresholds, $Q3 + 3IQR$ and $Q1 - 3IQR$, known as the outer limits. Typically, data points outside the internal limit are considered outliers, while outliers beyond the external limit deviate further from the real value and should be removed from the original data at the first instance. These outliers can significantly impact the data structure and disrupt the correlation between the feature and the score. On the other hand, data points within the external limit are also considered outliers, but their deviation is less severe compared to outliers beyond the external limit. Figure 2.7 illustrates the data cleaning process. $(Q1 - 1.5IQR, Q1 + 1.5IQR)$ and $(Q1 - 3IQR, Q1 + 3IQR)$ are two pairs of thresholds. If the data are in the pink region, the data are outliers.

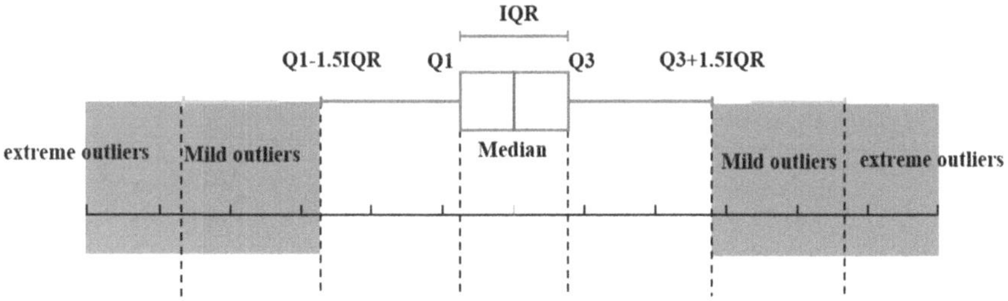

Fig. 2.7 Schematic diagram of the box plot. $(Q1 - 1.5IQR, Q1 + 1.5IQR)$ and $(Q1 - 3IQR, Q1 + 3IQR)$ represent two pairs of thresholds

2.1.4 Model Formulation

Deep learning [32, 50, 65] has emerged as a rapidly evolving research field. In a remarkable case, bionics principles have been applied to scientific research, and constructed mainly according to the human nervous system.

Architecture The basic unit of DNN is a neuron, and many of the neurons are interconnected to form the entire neural network. Our network architecture mainly includes seven layers. The first layer is the originally collected data, followed by a dimension reduction layer, a denoising layer, three hidden layers with nodes 12, 48, and 48, and an output layer with four nodes representing the four metrics. Each layer of neurons receives input from the previous neuron, and the output of a neuron undergoes two processes. The first process is to compute the weighted linear combination of inputs and the second process is to compute the nonlinear transformation of the results of the first process using the activation function. The network is divided into the forward process and the back propagation. The forward process computes the output from the first layer to the last layer, while the back propagation updates the parameters from the last layer to the first layer. Figure 2.8 illustrates the network architecture of the proposed model, where progressing from the first layer to the second layer is an FSS process, data cleaning occurs from the second layer to the third layer, the fourth, fifth, and sixth layers represent hidden layers, and the seventh layer is the output layer.

Forward propagation After finalizing the neural network structure, we commence training the network. During forward propagation, the output of each layer can be represented as:

$$\begin{aligned} v_n^{l+1} &= \sum_{m=1}^{N_{\text{node}}} \omega_{m,n} y_m^l + b_m^l \\ y_n^{l+1} &= \varphi(v_n^{l+1}) \end{aligned} \tag{2.4}$$

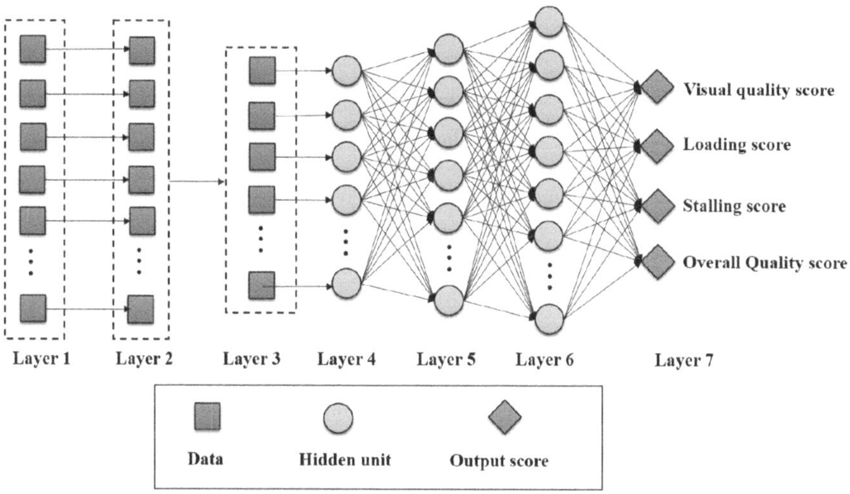

Fig. 2.8 Network structure of the proposed model. It consists of seven layers and is utilized to establish the relationships between the network parameters and the metric scores

where y_m^l denotes the mth node in the lth layer, $\omega_{m,n}$ represents the weight from the mth node to the nth node, b_m^l is the bias of the mth node in the lth layer, and $\varphi(\cdot)$ represents the nonlinear activation function [18]. The activation function plays a crucial role in transforming the linearity of inputs into nonlinearity, enabling a wide range of possibilities.

To differentiate between different scores, we design the loss function as:

$$L_{\text{loss}} = \sum_{x_{\text{sample}}} \log(1 + |y_{\text{gdv}} - y_{\text{est}}|) \tag{2.5}$$

where x_{sample} represents the samples, y_{gdv} is the ground truth values of the scores, and y_{est} is the estimated values of the scores.

Back propagation In the process of forward propagation, we compute the output scores from the first layer to the seventh layer. After the back propagation, we employ the gradient descent algorithm to minimize the loss function [10]. The network parameters are updated iteratively until the loss reaches a satisfactory level. The gradient update is performed as follows:

$$W^l = W^{l-1} - \beta \frac{\partial L_{\text{loss}}}{\partial W^l} \tag{2.6}$$

where β represents the learning rate, and $\frac{\partial L_{\text{loss}}}{\partial W^l}$ denotes the local gradient at the lth layer. For simplicity, the local gradient at node m is written as:

$$
\begin{aligned}
\Delta w_m &= \frac{\partial L_{\text{loss}}}{\partial y_m} \cdot \frac{\partial y_m}{\partial v_m} \cdot \frac{\partial v_m}{\partial w_m} \\
&= \frac{\partial L_{\text{loss}}}{\partial y_m} \cdot \varphi'(v_m) \cdot y_n
\end{aligned}
\tag{2.7}
$$

In order to solve (2.7), we need to compute $\frac{\partial L_{\text{loss}}}{\partial y_m}$ defined as:

$$
\begin{aligned}
\frac{\partial L_{\text{loss}}}{\partial y_m} &= \sum \frac{\partial L_{\text{loss}}}{\partial y_{\text{est}}} \cdot \frac{\partial y_{\text{est}}}{\partial v_{\text{est}}} \cdot \frac{\partial v_{\text{est}}}{\partial y_m} \\
&= \sum_k \frac{\partial L_{\text{loss}}}{\partial y_{\text{est}}} \cdot \varphi'(v_{\text{est}}) \cdot w_{kj}
\end{aligned}
\tag{2.8}
$$

The loss function is a piecewise function. According to (2.5), $\frac{\partial L_{\text{loss}}}{\partial y_{\text{est}}}$ can be written as:

$$
\frac{\partial L_{\text{loss}}}{\partial y_{\text{est}}} = \begin{cases} \sum \frac{-1}{1+y_{\text{gdv}}-y_{\text{est}}}, & \text{if } y_{\text{gdv}} > y_{\text{est}}, \\ \sum \frac{1}{1+y_{\text{est}}-y_{\text{gdv}}}, & \text{if } y_{\text{gdv}} < y_{\text{est}}. \end{cases}
\tag{2.9}
$$

According to (2.7)–(2.9), the local gradient at the hidden units is defined as:

$$
\begin{aligned}
\Delta w_m &= \sum \frac{\partial L_{\text{loss}}}{\partial y_{\text{est}}} \cdot \varphi'(v_{\text{est}}) \cdot w_{m,n} \cdot \varphi'(v_m) \cdot y_n \\
&= \sum_k \Delta w_k \cdot w_{m,n} \cdot \varphi'(v_m) \cdot y_n
\end{aligned}
\tag{2.10}
$$

where $\Delta w_{\text{out}} = \frac{\partial L_{\text{loss}}}{\partial y_{\text{est}}}$ is the local gradient at the output layer.

Utilizing neural networks for QoE prediction offers significant advantages. Particularly, it allows the derivation of conclusions that are not immediately evidence. Moreover, deep learning can reveal complex relationships between data and results. Notably, interactions between feature parameters can be learned through full connections of the hidden layers, enabling the generation of final QoE scores. Furthermore, we implement the network in TensorFlow, an open-source machine learning framework and train the model with the Adam optimizer. The learning rate is set to 0.0001, the iteration number is 30,000, and the batch size is 128.

2.1.5 *Result and Analysis*

This section presents a series of experimental results to validate the effectiveness of our approach in predicting QoE. We begin by conducting ablation experiments to demonstrate the results of each stage. Subsequently, we perform experiments on our collected dataset using our proposed method as well as several related QoE predicting approaches, such as the support vector model (SVM) [27, 56], adaptive neuro-fuzzy inference system (ANFIS) model [60], decision tree [55], and DeepQoE model. The compared algorithms are implemented using the same training and testing data as our proposed approach.

The experimental results are divided into three parts, namely the results of feature selection, data cleaning, and QoE prediction, since the proposed approach is divided into the same three parts. Each part aims to verify the effectiveness of its corresponding stage in our proposed approach. Specifically, the results of feature selection are used to evaluate the performance of the feature selection when using the FSS method. The results of data cleaning are used to evaluate the performance of the data cleaning when using the box plot method. The results of the QoE assessment are used to evaluate the performance of the QoE prediction when using the DNN model.

2.1.5.1 Results of Feature Selection

To provide clear results, we review the steps involved in FSS, which encompasses four stages. In the generation stage, we employ bidirectional search to generate the feature subset. We use Spearman's correlation coefficient as the evaluation function and set the threshold to 0.05 for the stopping criterion. If the correlation coefficient between two variables is less than 0.05, we consider the causality between the two variables to be very small. By applying this method, we identify 16 parameters out of the 89 recorded feature parameters that are strongly correlated with users' subjective scoring. These 16 parameters are user buffer time, buffer delay, bit rate, video size, video average rate, video play total time, screen resolution long, video peak rate, video all peak rate, light intensity, phone screen brightness, ping average rate, signal strength, video stalling number, video stalling total time, and video stalling duration proportion. Figure 2.9 displays the resulting parameter subset along with their corresponding correlation coefficients.

Figure 2.9 illustrates the correlation coefficients between the 16 feature parameters and the four users' subjective scores. A positive correlation coefficient indicates that the subjective score can be expressed as a monotone-increasing function of the corresponding feature. A higher coefficient value signifies a stronger correlation and a greater weight of that feature in predicting user QoE. The observations from Fig. 2.9 align with common sense and established understanding. For example, there is a strong positive correlation between the overall quality score and the bit rate, video size, and video average rate. Conversely, these features exhibit a strong negative correlation with the loading quality and stalling quality. The higher the clarity, the larger the video size, and the larger the amount of data that need to be buffered. When the network transmission rate is constant, the buffer time will be longer. The longer the total playtime, the longer the buffer time and the longer the video start time, which lower the scores for loading and stalling quality. Rebuffering

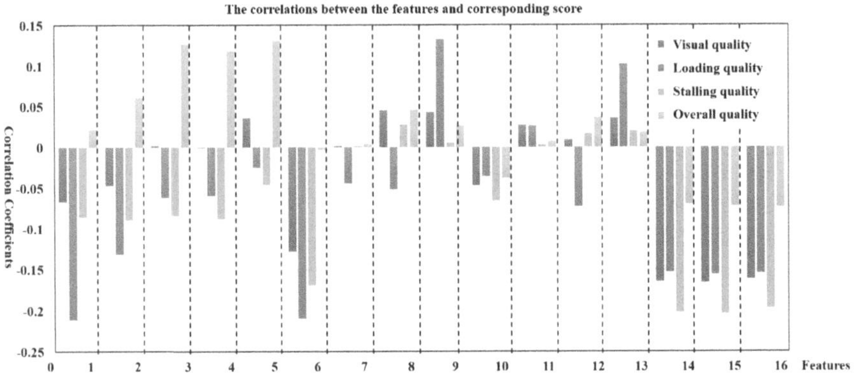

Fig. 2.9 Correlation coefficients between the features and the corresponding scores. The blue, red, green, and yellow bars represent the correlations between the features and the visual quality, loading quality, stalling quality, and overall quality, respectively. For each feature, we compute four correlations between the feature and its corresponding four scores. Features are labeled as 1–16: 1. user buffer time, 2. buffer delay, 3. bit rate, 4. video size, 5. video average rate, 6. video play total time, 7. screen resolution long, 8. video peak rate, 9. video all peak rate, 10. light intensity, 11. phone screen brightness, 12. ping average rate, 13. signal strength, 14. video stalling number, 15. video stalling total time, and 16. video stalling duration proportion. Moreover, a positive correlation indicates a positive relationship. The magnitude of the correlation coefficient reflects the strength of the correlation, whether positive or negative. The absolute value of the correlation coefficient does not exceed one

affects the user's viewing experience considerably, and users have low tolerance for rebuffering during video viewing. The number of video stalls, total stalling time, and stalling duration proportion notably affect the end users' scores, particularly in terms of visual quality, loading quality, and stalling quality.

2.1.5.2 Results of Data Cleaning

Since we collect data directly from users, the presence of outliers is inevitable. Consequently, we need to perform data cleaning on the selected database. In this part, we provide the data cleaning parameters and present the results. From the feature selection stage, it is evident that the selected 16 features exhibit strong correlations with the subjective scores.

From Fig. 2.9 and the rationality of the data, we identify four feature parameters for each score metric that exhibit higher correlation coefficients than the remaining 12 parameters. We select four parameters for each score metric as the standard for data cleaning as shown in Table 2.4. For visual quality, we select video play total time, user buffer time, video peak rate, and video average rate as they are most strongly correlated with visual quality. Similarly, we choose the four most correlated parameters for loading quality, stalling quality, and overall quality. Subsequently, we employ these selected parameters for data cleaning, enabling the removal of

Table 2.5 Model results of different scoring percentages. The best results were obtained when each of the five scores accounted for 20%

	5	4	3	2	1	Total RMSE
Samples	40%	30%	20%	8%	2%	2.5382
Samples	30%	30%	30%	5%	5%	1.0971
Samples	20%	20%	20%	20%	20%	0.3929

QoE metric is only related to a subset of these 89 network parameters. To identify the parameters associated with QoE, we compute the Pearson correlation coefficient between the collected parameters and the four metrics. Figure 2.9 represents the correlation coefficients between the features and the corresponding scores, displaying the 16 selected parameters with high correlation coefficients. Figure 2.10, on the other hand, showcases the correlation coefficients between the features and the corresponding scores after the data cleaning process. The selected 16 parameters remain the same as in Fig. 2.9. This confirms the correctness of the selected 16 parameters and further validates the performance of the data cleaning method.

2.1.5.3 Results of QoE Prediction

After conducting multiple attempts, we have determined the network architecture as shown in Fig. 2.8. Interestingly, we find that the QoE model achieves the best result when the number of nodes in the hidden layers is four times that of the first layer. Additionally, we select Relu $= \max(x, 0)$ as the activation function.

Once the structure of the neural network is decided, we conduct several experiments to guide us in selecting the data structures. We perform analyses and present the corresponding results in Table 2.5. Furthermore, we use root mean squared error MSE (RMSE) and mean absolute error (MAE) as measures to quantify the differences between the predicted QoE values and the true values. RMSE represents the square root of the second sample moment of the differences between predicted values and true values or the quadratic mean of these differences, while MAE measures the difference between two continuous variables.

$$\mathrm{RMSE} = \frac{\sqrt{\sum_{i=1}^{n}(x_i - \hat{x}_i)^2}}{n} \tag{2.11}$$

$$\mathrm{MAE} = \frac{|x_i - \hat{x}_i|}{n} \tag{2.12}$$

From Table 2.5, we observe that the model struggles to find the optimal solution for imbalanced samples. To address this issue, we conducted numerous experiments and find that the best results are obtained when the percentage of each score is equal. In other words, when the percentage of each fraction is 20%, the model achieves

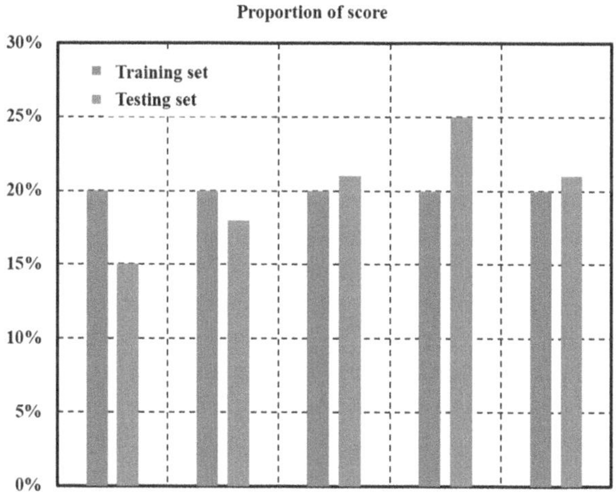

Fig. 2.11 Percentage of the training database and percentage of predicted scores in the test database

Table 2.6 Numerical results of different methods

Indexes	ANFIS1	ANFIS2	SVM	Decision tree	DeepQoE	The proposed approach
RMSE	1.3883	1.7386	1.8719	1.0315	1.0327	0.8686
MAE	1.1937	1.4048	0.6935	0.5837	0.5633	0.7609

the optimal prediction. In the test database, we also ensure that the percentage of each score is the same. The predicted scores for the test database are illustrated in Fig. 2.11, where the percentage of each score should be the same, i.e., each score should be 20%. However, there may be slight deviations from 20% for scores 1 and 4. Overall, our model demonstrates robust performance.

We compare our proposed method with ANFIS, SVM, decision tree, and DeepQoE, using RMSE and MAE metrics on the test set. For the ANFIS algorithm, we test two different versions referred to as ANFIS1 and ANFIS2. Our prediction model utilizes the 16 selected features. Through extensive network adjustments and structural optimization, we achieve the current performance level. From the results, it is evident that our proposed method achieves the smallest RMSE value among the compared methods on the dataset. The MAE value is comparable to that of the decision tree and DeepQoE methods, indicating that our approach performs favorably when compared to state-of-the-art methods. This demonstrates the powerful capability of DNN in capturing the complex relationship between network feature parameters and user scores. The numerical results are summarized in Table 2.6.

2.1.6 Conclusion

This chapter has introduced a large-scale dataset and a deep learning–based approach for data-driven objective QoE assessment in mobile video transmission. We developed the CU-vMOS mobile app to collect user QoE scores while viewing videos over the mobile Internet, enabling the establishment of a comprehensive QoE dataset in a real environment. The dataset includes four metric scores: visual quality, loading, stalling, and overall quality, along with 89 network parameters. Through data mining, we identified the specific network parameters correlated with each QoE score. Subsequently, we proposed a DNN structure for predicting subjective QoE scores based on the selected parameters. The experimental results demonstrated the capability of our large-scale QoE dataset to analyze the correlation between network parameters and users' QoE in mobile video transmission. Moreover, our proposed approach advanced the state-of-the-art objective QoE assessment for mobile video transmission, with high accuracy in predicting subjective QoE scores.

2.2 EEG-Based Image Perceptual Quality Assessment

With the proliferation of multimedia data, the evaluation of QoE has become crucial in multimedia transmission. Consequently, there is a growing need for QoE-oriented image quality assessment (IQA) methods. However, traditional user-based assessment methods are limited by deviations caused by human cognitive activities. In this chapter, we propose a brain-inspired IQA method based on EEG feature learning, which is a psychophysiological approach for studying human perception in IQA. Our method involves establishing an EEG dataset by collecting corresponding EEG signals, while subjects view distorted facial images. We then design a siamese network to extract EEG features that can distinguish image quality levels and measure user scores. The siamese network establishes the relationship between image quality and QoE as reflected by the EEG scores. The relationship is subsequently embedded into a prediction network that directly obtains EEG scores from images with different qualities. This enables the prediction of EEG scores through end-to-end learning. Experimental results demonstrate that our proposed method not only better evaluates the perceptual quality of facial images and reflects real human perceptions but also achieves superior score prediction performance on facial image datasets.

2.2.1 Introduction

In recent years, the evaluation of image and video quality in communication has gained significant importance due to the rise in multimedia services. The proper evaluation of multimedia communication systems can lead to further performance enhancements. In many scenarios, such as video conferences, images, and videos containing human faces as the primary content are important for communication.

Existing image quality assessment (IQA) methods can be categorized into objective and subjective quality assessments. The subjective quality assessment reflects human perception. Since image and video services are human-oriented, users' subjective experience is a very important evaluation index.

Traditional objective IQA methods, such as mean squared error (MSE) and peak signal-to-noise ratio (PSNR), evaluate image quality based on the differences between an original image and a distorted image. However, these objective methods have limitations. Images with high PSNR values may not necessarily result in a good user perception experience, as human visual perception is insensitive to certain distortions that affect PSNR. Structural similarity (SSIM) [71] can partially account for the characteristics of the human visual system, but it shares similar limitations as PSNR. In recent years, many IQA methods [29, 31, 41] based on machine learning can predict the quality scores of input images. However, these methods fail to make full use of human perceptual experiences when predicting objective indices.

Mean opinion score (MOS) [24] is a subjective assessment method based on user scoring. However, MOS is sensitive to deviations caused by human cognitive activities, as user scoring can be influenced by external factors. With advances in neuroscience, EEG is catching increasing attention in the fields of communications and signal processing. Studies [2–4] have explored the relationship between EEG responses and image quality degradation, employing methods such as steady-state visual evoked potential (SSVEP) and event-related potential (ERP) to reflect image and video qualities. EEG captures fundamental brain-level information and can serve as a complement to scoring. Mining and learning from human EEG data can reveal critical regions and reaction mechanisms associated with quality assessment. This opens up the possibility of utilizing EEG signals and these mechanisms to construct an assessment model that more accurately measures human perceptions. However, directly employing EEG signals to evaluate image quality is challenging due to the strict environmental requirements for EEG experiments, making it difficult to collect effective EEG signals under ordinary conditions. In this chapter, we combine EEG with machine learning methods to address this challenge.

Here, we propose a brain-inspired IQA method based on EEG feature learning for facial images in video conference scenarios. Our method consists of two components: an EEG feature extraction network and an EEG score prediction network. First, we collect corresponding EEG signals while subjects view distorted facial images. Subsequently, we design a siamese network to extract EEG features that can measure image quality scores, denoted as EEG scores. To directly output predicted EEG scores from images in practical scenarios, image features are extracted and mapped to EEG scores through a prediction network. The main contributions of this chapter can be summarized as follows:

- A quality degradation dataset is constructed by applying JPEG compression to images featuring human faces as the main targets, and the EEG dataset is collected from five subjects. Through the analysis of the EEG data, we verify that EEG features exhibit certain consistency across different facial images, while EEG signals corresponding to different degradation levels exhibit discernible differences.

- An EEG feature extraction network is constructed to extract features related to image quality degradation from single-trial EEG signals. These features are then mapped to EEG scores that reflect image quality. The validity of the EEG scores is verified through correlation analysis with other assessment indices.
- An EEG score prediction network is constructed to directly predict EEG-based quality scores from input images. This network extracts image features and maps them to image feature embeddings, from which the EEG scores can be calculated.

2.2.2 *Database Preparation*

This section provides a detailed description of the database preparation process, including the experiment materials, EEG experiment design, and data pretreatment.

2.2.2.1 Stimuli and Participants

Facial images are selected as the experiment materials, as the human face is a crucial element in video conference scenes. Due to the complexity of people's EEG responses to images, only one distortion-type JPEG is selected to control the variables. In this chapter, we apply JPEG compression to the Brazilian FEI facial image dataset [11] to obtain stimuli materials. The compression quality factor (Q) ranges from 1 to 100, resulting in 100 compressed images for each original image.

According to previous studies [14], human perception of images includes several just noticeable distortion thresholds. Therefore, we selected compressed images with Q values of 1, 12, 20, 28, 36, 52, and 100 as experiment stimuli. Five subjects, consisting of three males and two females aged 21–33, were invited to participate in the experiments. The experimental protocol was approved by the Medical Ethics Committee of the university.

2.2.2.2 Experiment Design

The experiments were carried out in an electromagnetic shielding room. A screen with a resolution of 1920×1080 pixels was used to present the image stimuli. According to [40], the subjects were seated at a distance of 90 cm from the screen. EEG signals were collected using a 64-channel EEG signal equipment (Brain Products, Germany) with a sampling rate of 500 Hz. EEG signals typically contain significant noise, necessitating multiple repetitions of the EEG experiments to achieve a high signal-to-noise ratio.

Prior to the formal experiments, a series of pre-experiments were conducted to familiarize the subjects with the experimental procedures. In the formal experiments, each trial consisted of four steps: a gray screen (B_1), the original facial image (O), a compressed image belonging to one of the seven compression levels (C), and another gray screen with a question (B_2). The gray screen B_1 was presented for 500 ms, while the original image O and the compressed image C were displayed for 1200 ms each. The gray screen with a question B_2 remained until the subject pressed keyboard buttons to answer the question: "Did you observe image distortion?" During the experiments, the subjects were instructed to fixate their gaze on the center of the screen. Each experiment session comprised ten parts, and in each part, images of the seven compression levels for five original images (35 trials in total) were randomly displayed once. Each subject completed one to four sessions.

2.2.2.3 EEG Data Pretreatment

EEG signals are characterized by low amplitude and often contain artifacts. Consequently, EEG signals can be considered a mixture of signals from various independent sources. By using the independent component analysis (ICA) method, the artifacts can be removed or mitigated. The EEG pretreatment process is as follows [13]:

1. Load EEG data and import scalp location information.
2. Perform re-referencing by averaging the potentials of electrodes TP9 and TP10.
3. Remove ineffective signals.
4. Apply a finite impulse response (FIR) filter (4–40 Hz) to process the EEG signals.
5. Epoch the EEG signals by selecting the EEG data from 300 ms before the onset of the target stimulus to 1200 ms after the onset as effective samples.
6. Apply ICA to the EEG signals.
7. Utilize the ADJUST plug-in [46] to remove eye movement signals and muscle movement signals.

After the pretreatment, ten EEG signal samples could be extracted for each subject from each compressed image in one experiment session. In total, 350 EEG signal samples could be obtained in one session, considering the seven compression levels of five original images.

2.2.3 *Proposed Method*

This section provides a detailed description of the proposed brain-inspired IQA method based on EEG feature learning. It includes an analysis of the EEG database and the structures of the proposed networks.

2.2.3.1 Database Analysis

Each channel's EEG signal for a subject represents a sequence of voltage values that change over time. Denote the EEG signal of a subject as $X_{i_s,j_c,n_t}(k,t)$, where i_s, j_c, and n_t represent the number of subjects, the compression level of the facial image, and the number of trials for each compression level, while k and t represent the channel number and time index of EEG data. The event-related potential (ERP) corresponding to compression level j_c for subject i_s can be expressed as follows:

$$\overline{X}_{i_s,j_c}(k,t) = \underset{k}{\text{mean}}\, X_{i_s,j_c,n_t}(k,t). \tag{2.13}$$

The average ERP value corresponding to compression level j_c or all subjects can then be calculated as follows:

$$\overline{X}_{j_c}(k,t) = \underset{i_s}{\text{mean}}\, \overline{X}_{i_s,j_c}(k,t). \tag{2.14}$$

As indicated in [38] and [4], ERP has been shown to reflect the brain's response to image quality degradation. Electrodes O_z, O_1, O_2, and CP_Z, associated with brain perception regions, were selected to calculate ERP for different image compression levels. In this method, the P300 value, a typical ERP occurring 300 ms after the stimulus, was used to verify the effectiveness of the EEG dataset.

In practical applications, acquiring EEG signals from subjects for multiple repetitions is challenging. Therefore, it is feasible to automatically extract features related to image quality from single-trial EEG signals. Machine learning algorithms can be employed to extract EEG features from single-trial EEG signals as the IQA score of the presented image.

2.2.3.2 Overview of the Network

In this chapter, we propose an IQA method based on EEG and machine learning. The overall structure of the method is illustrated in Fig. 2.12. Our goal is to develop an IQA network that can predict EEG scores from input images using human EEG signals. The proposed method consists of two subnetworks: the EEG feature extraction network and the EEG score prediction network. The EEG feature extraction network extracts EEG features from collected EEG signals and maps them to EEG scores, enabling the construction of an assessment standard for facial images based on EEG. On the other hand, the EEG score prediction network extracts image features and directly maps them to EEG scores.

In the EEG feature extraction network, facial images with different distortion levels are used as EEG experiment materials. First, the EEG signals are processed manually by the ERP analysis to verify the effectiveness of the dataset. Then, the proposed EEG feature extraction network automatically extracts features related to image quality from the EEG signals.

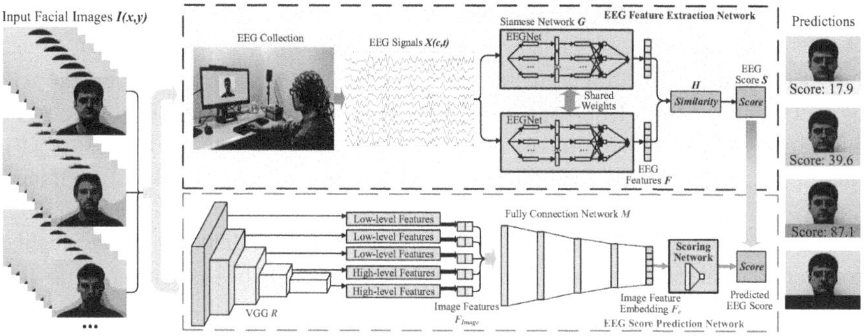

Fig. 2.12 Overall structure of the proposed brain-inspired IQA method with EEG feature learning

The EEG feature, F_{EEG}, is obtained by applying the feature extraction network W_G to the EEG signal $X(k, t)$ as follows:

$$F_{\mathrm{EEG}} = W_G(X(k, t)). \tag{2.15}$$

The EEG feature, F_{EEG}, is then mapped into the score space by distance calculation with normalization d_H. Consequently, the EEG score S_{EEG} can be obtained as:

$$S_{\mathrm{EEG}} = d_h(F) = d_H(W_G(X(k, t))). \tag{2.16}$$

For the EEG score prediction network, facial images with different distortion levels are input. The distortion-related information of the images is extracted using VGG-19, and the low-level and high-level features from VGG-19 are concatenated to form image features. A fully connected network is used to map the image features to image feature embeddings, which are ultimately mapped as the predicted EEG scores.

The feature F_{Image} of an input image $I(x, y)$ is extracted using VGG-19 R_{VGG} and then mapped to the image feature embedding F_e through the mapping network W_M. That is,

$$F_{\mathrm{Image}} = R_{\mathrm{VGG}}, \tag{2.17}$$

$$F_e = M(F_{\mathrm{Image}}) = W_M(R_{\mathrm{VGG}}(I(x, y))). \tag{2.18}$$

Consequently, EEG scores can be obtained from EEG features and can be predicted from image features through end-to-end learning.

2.2.3.3 EEG Feature Extraction Network

The EEG feature extraction network performs the EEG-to-score process by extracting EEG features from EEG signals and mapping them to EEG scores. In practical applications, it is challenging to collect EEG signals repeatedly from subjects. Therefore, an automated method is needed to extract features related to image quality.

The EEG feature extraction network extracts features related to image quality from single-trial EEG signals. The siamese network [51, 67] is employed as the basic structure, with each branch of the siamese network utilizing EEGNet [30] as the feature extraction network. EEGNet comprises convolutional layers, depthwise convolutional layers, and separable convolutional layers. The kernels used in EEGNet are illustrated in Fig. 2.13. The first block of EEGNet employs four convolutional kernels of size (1,250) to extract temporal features. The second block utilizes two depthwise kernels of size (61,1) to extract channel features for each feature map. The third block uses a separable convolution with depthwise kernels of size (1,16), followed by eight pointwise convolutions of size (1,1). Two EEG signals from the dataset are randomly selected to form a pair and are then input to the siamese network. The output of the network represents the similarity between the two EEG signals, measured by the difference among EEG features. If the EEG feature embedding corresponding to the original image is $F_{\text{EEG,original}}$, a smaller difference between an EEG feature embedding F_{EEG} and $F_{\text{EEG,original}}$ indicates better perceptual quality of the image corresponding to F_{EEG}, leading to a higher score. Therefore, the similarity of EEG signals can be used to assess image quality.

During the application process, EEG signal pairs are fed into the EEG feature extraction network. The training objective is to make the EEG features corresponding to the same distortion level become increasingly close, while making the EEG features of different distortion levels increasingly distant. Thus, the following contrastive loss is employed as the loss function:

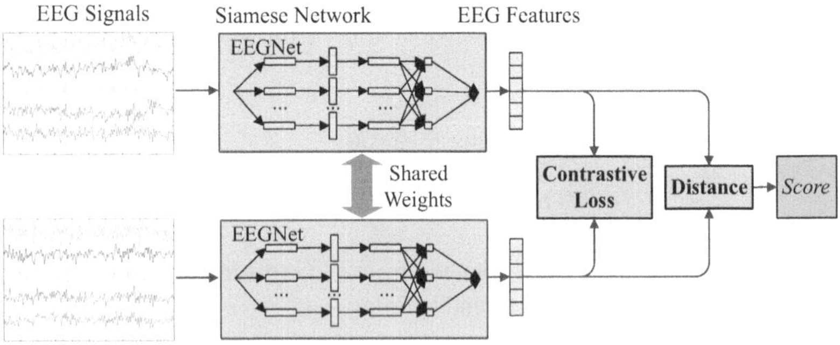

Fig. 2.13 Structure of the EEG feature extraction network

$$L_{\text{con}} = \frac{1}{2N_n} \sum_{n=1}^{N_n} (y_{\text{sample}} d_{\text{sample}}^2 + (1-y) \max(t_m - d_{\text{sample}}, 0)^2), \tag{2.19}$$

where d_{sample} represents the distance between two samples, t_m is the set threshold, and y_{sample} indicates whether the two samples match. $y_{\text{sample}} = 1$ indicates a match, while $y_{\text{sample}} = 0$ indicates a mismatch.

After training the network, each branch of the siamese network can output EEG feature embeddings. The distance between the EEG feature embedding corresponding to the distorted image and the undistorted image can be used to measure image quality. The cosine similarity is chosen as the measure of difference or similarity between EEG signals:

$$D_{\text{cos}} = \frac{\sum_{i=1}^{n}(a_i \times b_i)}{\sqrt{\sum_{i=1}^{n}(a_i)^2} \times \sqrt{\sum_{i=1}^{n}(b_i)^2}}, \tag{2.20}$$

where $\mathbf{a} = [a_1, a_2, \ldots, a_n]$ and $\mathbf{b} = [b_1, b_2, \ldots, b_n]$ represent two n-dimensional vectors.

After normalization, the similarity between the EEG feature embedding corresponding to the distorted image and the undistorted image can be used as the image quality score, defined as the EEG score. Since EEG signals contain information related to human perception of image quality, this measurement method can reflect real human perception of distortion to a certain extent.

2.2.3.4 EEG Score Prediction Network

The purpose of the EEG score prediction network is to complete the image-to-score process, obtaining EEG scores from facial images. The network extracts image features from the facial images, maps the image features to image feature embeddings, and finally calculates EEG scores. To utilize EEG scores in practical situations, it is necessary to bypass the EEG recording stage and directly extract features from the images. The VGG-19 network is employed to extract image features, and the feature maps of each block of VGG are concatenated after dimensionality reduction by 1×1 convolution and pooling to obtain image features. Four fully connected (FC) layers with 4096, 4096, 1000, and 48 neurons map the image features to feature embeddings. After passing through the scoring network with two FC layers containing 25 and 1 neurons, the EEG scores can be calculated.

During the application process, facial images with different distortion levels are input. The loss function is defined as follows:

$$L_{\text{score}} = \frac{1}{2N_n} \sum_i ||S_p - S_t||_2^2 \tag{2.21}$$

where S_p represents the EEG scores predicted by the proposed network and S_t represents the EEG scores obtained by the EEG feature extraction network.

2.2.4 Experiment Results and Discussions

In this section, we present the experiment results to demonstrate the performance of assessing the quality of facial images and compare them with other methods to verify the effectiveness of the proposed method.

2.2.4.1 Data Analysis

(1) Behavioral Data The subjects were asked to indicate whether they observed image quality degradation by pressing buttons. Therefore, the behavior data of the subjects were obtained as an index of the detection rate.

(2) EEG Data The EEG data underwent pretreatment and ERP analysis according to the methods mentioned in Sect. 2.2.2. P300 values, with the maximum amplitudes between 200 and 600 ms, are obtained as shown in Fig. 2.14. From Fig. 2.14, the P300 values of the EEG signals corresponding to different image quality levels can be taken as an index of image quality and considered as a measurement of QoE. It can be seen from Fig. 2.14 that different image quality levels can stimulate EEG signals with different shapes. The scalp topographic maps during view of images with different quality levels are shown in Figs. 2.15 and 2.16. When images with

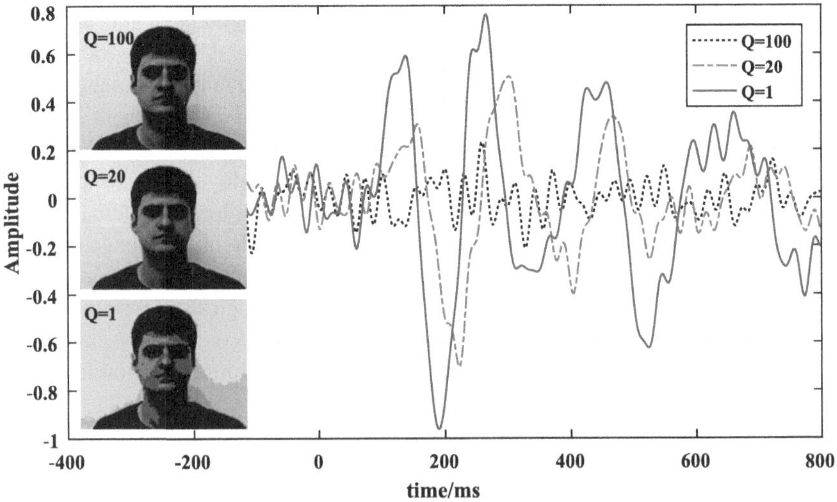

Fig. 2.14 ERP results of the EEG signals

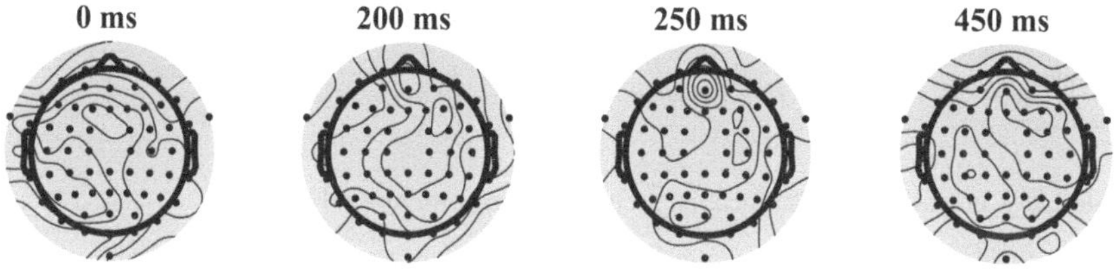

Fig. 2.15 Scalp topographic maps generated during view of undistorted images

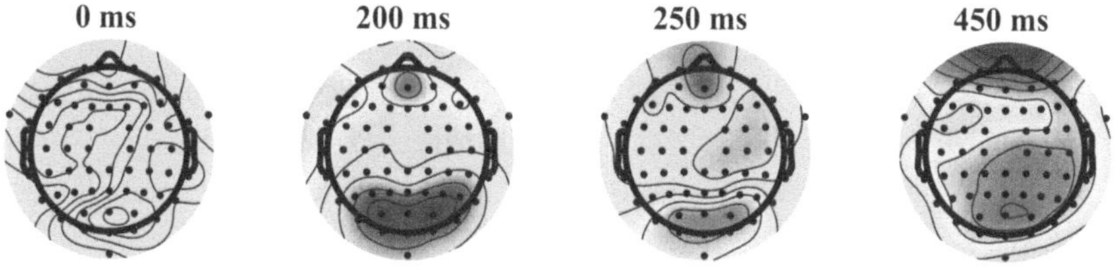

Fig. 2.16 Scalp topographic maps generated during view of distorted images

different distortion levels are used as stimuli, the brain produces different activities. These results show that the EEG signals collected in this experiment are effective and confirm the feasibility of using EEG signals as a measurement of image quality.

2.2.4.2 Evaluation of the EEG Feature Extraction Network

(1) EEG Score Each siamese network outputs an EEG feature embedding with a dimension of 48. The cosine similarity after normalization is used to measure the difference between the EEG signal corresponding to the image to be evaluated and the EEG signal corresponding to the undistorted image. This difference represents the score of the image. Five distortion levels ($Q = 1, 12, 20, 28, 100$) are selected as key distortion levels, resulting in 250–1000 EEG samples for each subject. The data are divided into an 80% training set, 10% validation set, and 10% test set. The accuracy of the siamese network on the test set is 72.4%, indicating that the EEGNet-based siamese network effectively extracts EEG features. EEG scores corresponding to these five distortion levels are obtained by averaging the test sets. To obtain a more continuous curve of the EEG scores with different distortion levels, a fitting method is employed.

(2) Classification Verification The performance and effectiveness of the EEG feature extraction network are evaluated through a classification task. The LSTM [51], EEGNet [30], and our proposed method are used to classify EEG signals with different distortion levels. As shown in Table 2.7, the classification accuracy rates of these three methods on the test set are 47.6%, 69.2%, and 72.4%, respectively.

Table 2.7 Comparison of classification accuracy among different methods

Methods	LSTM [51]	EEGNet [30]	Our method
Accuracy	47.6%	69.2%	**72.4%**

Table 2.8 Comparison of classification accuracy among different subjects

Sub.	1	2	3	4	5	All
Acc.	77.0%	68.0%	72.0%	78.6%	64.0%	72.4%

Table 2.9 Correlation comparison with some assessment standards

Index		P300 values	MOS
BRISQUE [44]	PLCC	0.9835	0.9587
	SRCC	0.9720	0.9485
ILNIQE [69]	PLCC	0.9833	0.9609
	SRCC	0.9565	0.9463
SUR	PLCC	0.9913	0.9006
	SRCC	0.8737	0.8447
Ours	PLCC	**0.9973**	**0.9769**
	SRCC	**0.9919**	**0.9549**

The classification accuracy results of the proposed method on datasets of each subject are shown in Table 2.8. Due to individual differences, the model performs differently on different subjects' datasets. However, the model is able to learn common features related to quality perception among subjects and perform well on the dataset consisting of all subjects' data.

(3) Correlation with Other Assessment Standards To verify the effectiveness of the EEG score standard, we calculate the correlation between the EEG score and P300 obtained through the ERP method mentioned in [4] and MOS. Pearson's linear correlation coefficient (PLCC) and Spearman's rank order correlation coefficient (SRCC) are calculated. The results are shown in Table 2.9. Compared with BRISQUE, ILNIQE, and satisfied user ratio (SUR) standards, the EEG score extracted by the network from EEG signals shows a strong correlation with P300 and MOS. This indicates that the EEG score is effective in assessing image perceptual quality. It is worth mentioning that SUR and EEG score here are predicted using SUR-Net [14] and our prediction network, respectively.

2.2.4.3 Evaluation of the EEG Score Prediction Network

(1) Correlation Analysis The EEG score prediction network can directly predict EEG scores from images, as shown in Fig. 2.17. The EEG scores obtained by the EEG feature extraction network can be used as the ground truth. The correlation between the EEG score predicted by the network and the ground truth can be calculated to verify the effectiveness of the network. The results are shown in

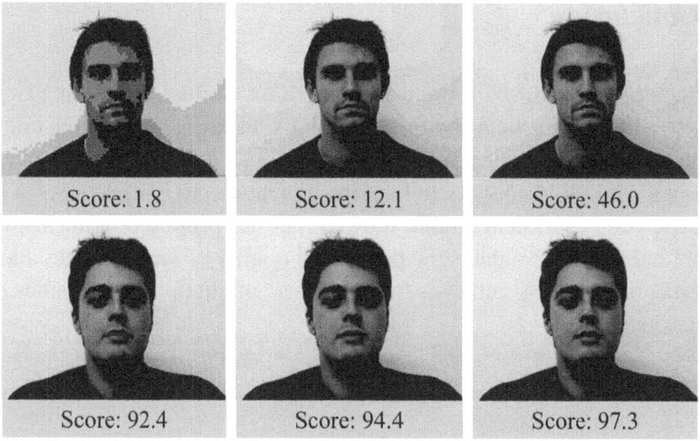

Fig. 2.17 Prediction results of the proposed EEG score prediction network

Table 2.10 Comparison with other prediction methods

Methods		FEI dataset	IMM dataset
Hyper Net [59]	PLCC	0.9994	0.9964
	SRCC	0.9900	0.9371
SUR-Net [14]	PLCC	0.9971	0.9890
	SRCC	0.9844	0.9214
ResNet [20]	PLCC	0.9932	0.9841
	SRCC	0.9789	0.9153
VGG-19 [57]	PLCC	0.9912	0.9824
	SRCC	0.9773	0.9147
Our method	PLCC	*0.9992*	**0.9973**
	SRCC	**0.9990**	**0.9845**

Table 2.10. From the table, it can be observed that the predicted EEG score and the ground truth have a strong correlation, indicating the effectiveness of the prediction network.

(2) Comparison with Other Methods We also use the adaptive hypernet [59], SUR-Net [14], ResNet, and VGG methods to predict scores, and the results are shown in Table 2.10. The method in [59] adapts well to different image contents and predicts the MOS score of different kinds of images. Since the IQA method proposed in this chapter focuses on the assessment of facial images, where the difference among images is small, the score is less affected by the content. From the results in Table 2.10, it can be seen that the score prediction network proposed in this chapter achieves comparable results with the state-of-the-art score prediction network.

(3) Generalization Ability Test In addition to the FEI dataset, we also used the IMM face datasets [58] for verification, and our method demonstrates a better performance, as shown in Table 2.10.

2.2.5 Conclusion

In this chapter, we proposed a brain-inspired IQA method based on EEG feature learning to address practical problems in IQA tasks. The method combines a psychophysiological approach to measure QoE and reflect human perception of images, and a prediction network to map image features to EEG scores for practical applications. The experiment results showed that our proposed brain-inspired IQA method not only better evaluates the perceptual quality of facial images and reflects human perception but also achieves superior score prediction performance on facial image datasets.

References

1. Andrews, M., Kumaran, K., Ramanan, K., Stolyar, A., Whiting, P., Vijayakumar, R.: Providing quality of service over a shared wireless link. IEEE Commun. Mag. **39**(2), 150–154 (2001)
2. Arndt, S., Antons, J., Schleicher, R., Moller, S., Scholler, S., Curio, G.: A physiological approach to determine video quality. In: 2011 IEEE International Symposium on Multimedia, pp. 518–523 (2011). https://doi.org/10.1109/ISM.2011.91
3. Arndt, S., Antons, J., Schleicher, R., Moller, S., Curio, G.: Perception of low-quality videos analyzed by means of electroencephalography. In: 2012 4th International Workshop Quality of Multimedia Experience, pp. 284–289 (2012). https://doi.org/10.1109/QoMEX.2012.6263836
4. Arndt, S., Antons, J., Schleicher, R., Moller, S., Curio, G.: Using electroencephalography to measure perceived video quality. IEEE J. Sel. Topics Signal Process. **8**(3), 366–376 (2014). https://doi.org/10.1109/JSTSP.2014.2313026
5. Balachandran, A., Sekar, V., Akella, A., Seshan, S., Stoica, I., Zhang, H.: Developing a predictive model of quality of experience for internet video. ACM Sigcomm. Comput. Commun. Rev. **43**(4), 339–350 (2013)
6. BT.500: Methodology for the subjective assessment of the quality of television pictures. Int. Telecommun. Union **4**, 1–44 (2002)
7. Chen, S., Nahrstedt, K.: Distributed quality-of-service routing in ad hoc networks. IEEE J. Sel. Areas Commun. **17**(8), 1488–1505 (1999)
8. Chen, Y., Wu, K., Zhang, Q.: From QoS to QoE: a tutorial on video quality assessment. IEEE Commun. Surv. Tutor. **17**(2), 1126–1165 (2015)
9. Chikkerur, S., Sundaram, V., Reisslein, M., Karam, L.J.: Objective video quality assessment methods: a classification, review, and performance comparison. IEEE Trans. Broadcast. **57**(2), 165–182 (2011)
10. Cun, Y.L., Boser, B., Denker, J.S., Howard, R.E., Habbard, W., Jackel, L.D., Henderson, D.: Handwritten digit recognition with a back-propagation network. Adv. Neural Inf. Process. Syst. **2**(2), 396–404 (1989)
11. de Oliveira Junior, L.L., Thomaz., C.E.: Captura e alinhamento de imagens: Um banco de faces brasileiro. Undergraduate Technical Report (2006)
12. De Winter, J.C.F., Gosling, S.D., Potter, J.: Comparing the pearson and spearman correlation coefficients across distributions and sample sizes: a tutorial using simulations and empirical data. Psychol. Methods **21**(3), 273 (2016)
13. Duan, Y., Li, Z., Tao, X., Li, Q., Hu, S., Lu, J.: EEG-based maritime object detection for IoT-driven surveillance systems in smart ocean. IEEE Int. Things **7**(10), 9678–9687 (2020)
14. Fan, C., Lin, H., Hosu, V., Zhang, Y., Saupe, D.: SUR-Net: Predicting the satisfied user ratio curve for image compression with deep learning. In: 2019 International Conference on Quality of Multimedia Experience (2019)

15. Fiedler, M., Hossfeld, T., Tran-Gia, P.: A generic quantitative relationship between quality of experience and quality of service. IEEE Netw. **24**, 36–41. IEEE Press (2010)
16. Fodor, I.K.: A survey of dimension reduction techniques. Neoplasia **7**(5), 475–485 (2002)
17. Gadaleta, M., Chiariotti, F., Rossi, M., Zanella, A.: D-dash: a deep Q-learning framework for dash video streaming. IEEE Trans. Cognit. Commun. Netw. **PP**(99), 1–1 (2017)
18. Glorot, X., Bordes, A., Bengio, Y.: Deep sparse rectifier neural networks. In: International Conference on Artificial Intelligence and Statistics, pp. 315–323 (2011)
19. Group, T.V.Q.E.: Final report from the video quality experts group on the validation of objective models of video quality assessment (2000)
20. He, K., Zhang, X., Ren, S., Sun, J.: Deep residual learning for image recognition. In: Proceedings of the IEEE Conference on Computer Vision and Pattern Recognition (2016)
21. He, X., Wang, K., Huang, H., Miyazaki, T., Wang, Y., Guo, S.: Green resource allocation based on deep reinforcement learning in content-centric iot. IEEE Trans. Emerg. Topics Comput. **PP**(99), 1–1 (2018)
22. Holte, R.C., Felner, A., Sharon, G., Sturtevant, N.R.: Bidirectional search that is guaranteed to meet in the middle. In: Thirtieth AAAI Conference on Artificial Intelligence, pp. 3411–3417 (2016)
23. Hou, W., Gao, X., Tao, D., Li, X.: Blind image quality assessment via deep learning. IEEE Trans. Neur. Netw. Learn. Syst. **26**(6), 1275–1286 (2017)
24. ITU-T P.800: Methods for subjective determination of transmission quality. ITU-T Recommendation P.800 (1996)
25. Jain, R: Quality of experience. Multimedia IEEE **11**(1), 96–95 (2010)
26. Jian, H., Wen, Y., Huang, J., Di, W.: On the cost–QoE tradeoff for cloud-based video streaming under amazon ec2's pricing models. IEEE Trans. Circ. Syst. Video Technol. **24**(4), 669–680 (2014)
27. Joachims, T.: Making large-scale svm learining practical. Adv. Kernel Methods Suppo. Vector Learn. **8**(3), 499–526 (2006)
28. Kim, J., Lee, S.: Deep learning of human visual sensitivity in image quality assessment framework. In: IEEE Conference on Computer Vision and Pattern Recognition, pp. 1969–1977 (2017)
29. Kim, J., Nguyen, A., Lee, S.: Deep CNN-based blind image quality predictor. IEEE Trans. Neural Netw. Learn. Syst. **30**(1), 11–24 (2019). https://doi.org/10.1109/TNNLS.2018.2829819
30. Lawhern, V.J., Solon, A.J., Waytowich, N.R., Gordon, S.M., Hung, C.P., Lance, B.J.: Eegnet: A compact convolutional network for EEG-based brain-computer interfaces. J. Neural Eng. **15**(5), 056013.1–056013.17 (2016)
31. Le, K., Peng, Y., Yi, L., Doermann, D.: Convolutional neural networks for no-reference image quality assessment. In: 2014 IEEE Conference on Computer Vision and Pattern Recognition (2014)
32. Lecun, Y., Bengio, Y., Hinton, G.: Deep learning. Nature **521**(7553), 436 (2015)
33. Li, X., Guo, Q., Lu, X.: Spatiotemporal statistics for video quality assessment. IEEE Trans. Image Process. **25**(7), 3329–3342 (2016)
34. Li, Y., Jin, D., Wang, Z., Hui, P., Zeng, L., Chen, S.: A markov jump process model for urban vehicular mobility: Modeling and applications. IEEE Trans. Mobile Comput. **13**(9), 1911–1926 (2014)
35. Li, Y., Wu, T., Hui, P., Jin, D., Chen, S.: Social-aware D2D communications: qualitative insights and quantitative analysis. IEEE Commun. Mag. **52**(6), 150–158 (2014)
36. Liang, Z., Zhen, Y., Wen, Y., Wang, H., Guizani, M.: Resource allocation with incomplete information for QoE-driven multimedia communications. IEEE Trans. Wirel. Commun. **12**(8), 3733–3745 (2013)
37. Lin, X., Ma, H., Luo, L., Chen, Y.: No-reference video quality assessment in the compressed domain. IEEE Trans. Consumer Electr. **58**(2), 505–512 (2012)
38. Lindemann, L., Magnor, M.: Assessing the quality of compressed images using EEG. In: 2011 18th IEEE International Conference on Image Processing, pp. 3109–3112 (2011). https://doi.org/10.1109/ICIP.2011.6116324

39. Liu, L., Sheng, C., Kai, L., Cheng, T.: Bidirectional search antenna selection algorithm in massive MIMO system. J. Beijing Jiaotong Univer. **40**(05), 56–62 (2016)
40. Liu, X., Tao, X., Xu, M., Zhan, Y., Lu, J.: An EEG-based study on perception of video distortion under various content motion conditions. IEEE Trans. Multimedia **22**(4), 949–960 (2020)
41. Ma, K., Liu, W., Zhang, K., Duanmu, Z., Wang, Z., Zuo, W.: End-to-end blind image quality assessment using deep neural networks. IEEE Trans. Image Process. **27**(3), 1202–1213 (2018). https://doi.org/10.1109/TIP.2017.2774045
42. Mao, H., Netravali, R., Alizadeh, M.: Neural adaptive video streaming with pensieve. In: Conference of the ACM Special Interest Group on Data Communication, pp. 197–210 (2017)
43. Mitra, K., Ahlund, C., Zaslavsky, A.: A decision-theoretic approach for quality-of-experience measurement and prediction. In: IEEE International Conference on Multimedia and Expo, pp. 1–4 (2011)
44. Mittal, A., Moorthy, A.K., Bovik, A.C.: No-reference image quality assessment in the spatial domain. IEEE Trans. Image Process. **21**(12), 4695–4708 (2012). https://doi.org/10.1109/TIP.2012.2214050
45. Mocanu, D.C., Exarchakos, G., Liotta, A.: Deep learning for objective quality assessment of 3d images. In: IEEE International Conference on Image Processing, pp. 758–762 (2015)
46. Mognon, A., Jovicich, J., Bruzzone, L., Buiatti, M.: Adjust: An automatic EEG artifact detector based on the joint use of spatial and temporal features. Psychophysiology **48**(2), 229–240 (2011)
47. Moorthy, A.K., Seshadrinathan, K., Soundararajan, R., Bovik, A.C.: Wireless video quality assessment: a study of subjective scores and objective algorithms. IEEE Trans. Circ. Syst. Video Technol. **20**(4), 587–599 (2010)
48. Moradi, P., Gholampour, M.: A hybrid particle swarm optimization for feature subset selection by integrating a novel local search strategy. Appl. Soft Comput. **43**(C), 117–130 (2016)
49. Narendra, P.M., Fukunaga, K.: A branch and bound algorithm for feature subset selection. IEEE Trans. Comput. **C-26**(9), 917–922 (2006)
50. O'Shea, T.J., Erpek, T., Clancy, T.C.: Deep learning based MIMO communications. arXiv:1707.07980 (2017)
51. Palazzo, S., Spampinato, C., Kavasidis, I., Giordano, D., Schmidt, J., Shah, M.: Decoding brain representations by multimodal learning of neural activity and visual features. IEEE Trans. Pattern Anal. Mach. Intell. 43, 1–1 (2020). https://doi.org/10.1109/TPAMI.2020.2995909
52. Pearson, E.: The test of significance for the correlation coefficient. Public. Amer. Statist. Assoc. **26**(174), 128–134 (1932)
53. Qiao, J., He, Y., Shen, X.S.: Improving video streaming quality in 5g enabled vehicular networks. IEEE Wirel. Commun. **25**(2), 133–139 (2018)
54. Rousseeuw, P., IdaRuts, Tukey, J.: The bagplot: a bivariate boxplot. Amer. Statist. **53**(4), 382–387 (1999)
55. Schapire, R.E., Freund, Y.: Boosting: Foundations and Algorithms. The MIT Press, Cambridge (2012). https://doi.org/10.7551/mitpress/8291.001.0001
56. Schuldt, C.: Recognizing human motions: A local svm approach. In: Proceedings of the IEEE International Conference on Pattern Recognition, vol. 3 (2004)
57. Simonyan, K., Zisserman, A.: Very deep convolutional networks for large-scale image recognition. arXiv:1409.1556 (2014)
58. Stegmann, M.B., Ersbøll, B.K., Larsen, R.: FAME—a flexible appearance modelling environment. IEEE Trans. Med. Imag. **22**(10), 1319–1331 (2003)
59. Su, S., Yan, Q., Zhu, Y., Zhang, C., Ge, X., Sun, J., Zhang, Y.: Blindly assess image quality in the wild guided by a self-adaptive hyper network. In: 2020 IEEE Conference on Computer Vision and Pattern Recognition, pp. 3664–3673 (2020). https://doi.org/10.1109/CVPR42600.2020.00372
60. Vasileva-Stojanovska, T., Vasileva, M., Malinovski, T., Trajkovik, V.: An anfis model of quality of experience prediction in education. Appl. Soft Comput. **34**(C), 129–138 (2015)

61. Vega, M.T., Mocanu, D.C., Famaey, J., Stavrou, S., Liotta, A.: Deep learning for quality assessment in live video streaming. IEEE Signal Process. Lett. **24**(6), 736–740 (2017)
62. Wang, Z., Crowcroft, J.: Quality-of-service routing for supporting multimedia applications. IEEE J. Sel. Areas Commun. **14**(7), 1228–1234 (2002)
63. Wang, Z., Sheikh, H.R., Bovik, A.C.: Objective video quality assessment. Handbook Video Databases Design Appl. **17**(5), 1041–1078 (2003)
64. Yang, F., Song, J., Wan, S., Wu, H.R.: Content-adaptive packet-layer model for quality assessment of networked video services. IEEE J. Sel. Topics Signal Process. **6**(6), 672–683 (2012)
65. Ye, H., Li, G.Y., Juang, B.H.F.: Deep reinforcement learning based resource allocation for V2V communications. IEEE Trans. Vehic. Technol. **68**(4), 3163–3173 (2019). https://doi.org/10.1109/TVT.2019.2897134
66. Yetgin, Z., Gocer, Z.: Quality of experience prediction model for progressive downloading over mobile broadcast networks. Telecommun. Syst. **58**(1), 1–12 (2015)
67. Zagoruyko, S., Komodakis, N.: Learning to compare image patches via convolutional neural networks. In: 2015 IEEE Conference on Computer Vision and Pattern Recognition, pp. 4353–4361 (2015). https://doi.org/10.1109/CVPR.2015.7299064
68. Zhai, G., Cai, J., Lin, W., Yang, X., Zhang, W.: Three dimensional scalable video adaptation via user-end perceptual quality assessment. IEEE Trans. Broadcast. **54**(3), 719–727 (2008)
69. Zhang, L., Zhang, L., Bovik, A.C.: A feature-enriched completely blind image quality evaluator. IEEE Trans. Image Process. **24**(8), 2579–2591 (2015). https://doi.org/10.1109/TIP.2015.2426416
70. Zhou, H., Deng, Z., Xia, Y., Fu, M.: A new sampling method in particle filter based on pearson correlation coefficient. Neurocomputing **216**, 208–215 (2016)
71. Wang, Z., Bovik, A.C., Sheikh, H.R., Simoncelli, E.P.: Image quality assessment: from error visibility to structural similarity. IEEE Trans. Image Process. **13**(4), 600–612 (2004). https://doi.org/10.1109/TIP.2003.819861

Chapter 3
Prior Knowledge Base

This chapter introduces the proposed framework for semantic communications, which includes semantic representations, semantic encoding and decoding, and semantic information transmission with the next-generation multiple access (NGMA) [28], as shown in Fig. 3.1. Specifically, rules, methods, models, and training libraries are deployed at the cloud server and maintained at the edge server to provide prior semantic knowledge for the transmitter and receiver. Based on the semantic knowledge base, the transmitter performs semantic representation and semantic encoding of the multimodal source. The abstracted semantic information from multiple users is then transmitted using multiple access techniques. The information can be transmitted using traditional methods such as bit sequences or the semantic transmission system to effectively transmit semantic features. At the receiver, with the aid of the semantic knowledge base, a generative approach is adopted to achieve semantic fidelity in reconstructing high-resolution, stable, and smooth video data.

3.1 Framework of Hierarchical Semantic Knowledge Base

The presentation of data in multimedia communications is becoming increasingly diverse, with an explosion of data in different modalities such as text, images, and videos. In this study, we explore a new model of content-oriented semantic representation by drawing inspiration from the cognitive mechanisms of the human brain. Specifically, we investigate a cognitive model–based digital knowledge modeling and representation method that imitates the cognitive and information storage properties of the human brain using neurons. To establish a general knowledge–level representation for data in different spaces, times, and modalities, we propose a spatiotemporal object-attribute-relation (OAR) model, as depicted in Fig. 3.2.

X. Tao et al., *Wireless Multimedia Computational Communications*,
Wireless Networks, https://doi.org/10.1007/978-3-031-64155-8_3

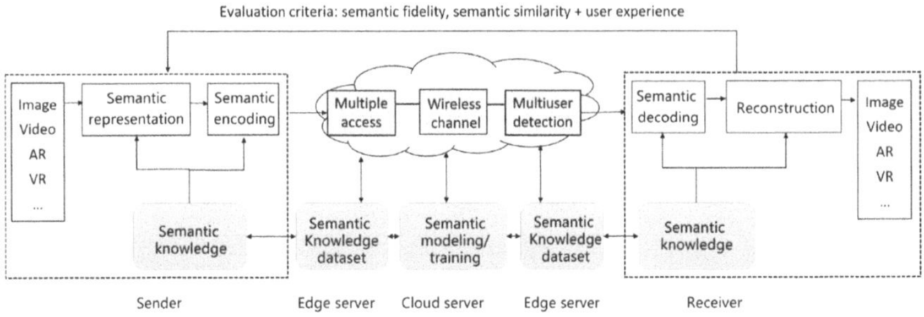

Fig. 3.1 Proposed semantic communication framework

Fig. 3.2 General OAR semantic representation

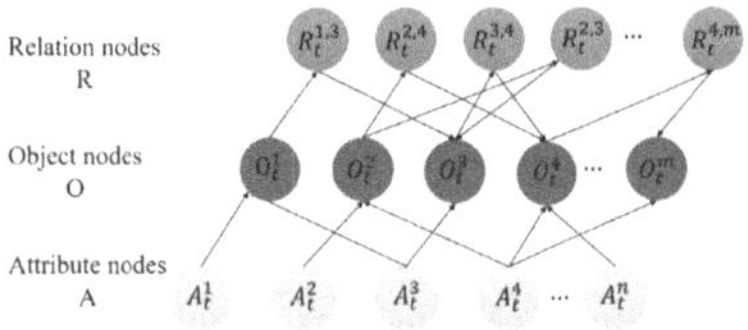

The OAR model employs objects and attributes as a generic representation of a concept and then specializes the concept in a specific scenario by establishing specific relationships, particularly between objects, to form a concept-specific representation. This concept-specific representation is generated based on semantics and incorporates knowledge. By utilizing relations, the OAR model achieves a mechanism that simulates the human brain's processing and storage of information. It extends the storage of knowledge beyond concepts themselves to the relationships between two concepts, resulting in a more accurate and comprehensive knowledge representation. Sketch graphs are primarily oriented toward human-oriented communication tasks, such as video surveillance and talking head conferences, while the OAR model is primarily oriented toward task-oriented machine communication, such as communication between intelligent bodies and holographic projection.

3.2 OAR-Based Semantic Knowledge Base

Event prediction involves analyzing and predicting events that occur at specific times and locations in order to take subsequent actions and make informed decisions. Current event prediction approaches mainly employ deep learning methods to analyze regular patterns from a large amount of historical data. Adversarial soccer event prediction remains a significant challenge due to its inherent antagonism and complex relationships between players. Taking this into consideration, we propose an OAR network for predicting soccer events using multimodal data, including spatiotemporal trajectory data and video data. The proposed scheme transforms multimodal data into an OAR space that integrates global and local relationships (adversarial information and multi-objective information) to improve prediction performance. Specifically, this scheme consists of three main components: a relation module, an object-attribute module, and a graph prediction module. First, we employ convolutional long short-term memory (ConvLSTM) to extract visual features of players from video data and use LSTM to extract movement features of players from spatiotemporal data. Then, we apply a multihead gated recurrent unit (GRU) attention mechanism to calculate relation weights. These three components are combined into an OAR graph representing a clip in a soccer game. Finally, an OAR graph neural network is designed to reason about the influence among different objects and predict events. The entire process forms an end-to-end event prediction learning framework. Extensive experimental results on two challenging datasets, soccER and Skill-Corner, verify the effectiveness of the proposed framework.

3.2.1 Introduction

Event prediction involves analyzing and predicting events that occur at specific times and locations in order to make informed actions and decisions. Event analytics and prediction have become widely used technologies in various fields such as healthcare, business, cybersphere, politics, sports, and entertainment, influencing almost every corner of our lives. These technologies provide valuable information for effective decision-making. Consequently, the analysis and prediction of events have attracted significant attention in recent years. Current event prediction approaches primarily rely on deep learning methods to analyze regular patterns from large amounts of historical data. One problem of existing schemes is that they only consider one or two directly related objects when predicting events, hence ignoring the influence of other indirectly related objects and focusing on only one modality of data: spatiotemporal data. Adversarial soccer event prediction remains a major challenge due to the strong antagonism and complex relationships between players.

In recent years, with the advent of deep learning techniques, various institutions and industrial corporations have begun to collect large amounts of data during games and training and use AI techniques for analysis and game result prediction, tactic development, sports game development, athlete injury management, and more [2]. Soccer, being one of the most popular and complex sports, presents numerous opportunities for applying event recognition and prediction. Analyzing soccer events, such as passes and shots, can help coaches analyze player and team tactics, develop AI tools for soccer games, and enhance game broadcasts by automatically scheduling cameras and highlighting key areas on the broadcast screen for adaptive bandwidth allocation. Soccer is characterized by the need for multiple players and the existence of cooperation and confrontation, making it insufficient to analyze solely based on historical information, such as players' and teams' past performance. Visual information and spatiotemporal trajectory information have been considered in some existing research, where visual data enable the use of identification of player posture and behavior, while spatiotemporal trajectory information allows the analysis of events [10]. Combining information from both modalities can lead to improved performance in event identification and prediction.

Traditional methods for event recognition rely on manually designed event features or simply classify individual player trajectories and images [18, 19, 24, 30]. For example, recognizing a pass can be broken down into two subevents: "the ball leaving a player's foot" and "a player receiving the ball." The features of these subevents can be obtained by analyzing the player's posture and the relative trajectory between the player and the ball. However, this approach can only identify completed events and does not predict events that have not yet started or finished, as their characteristics cannot be simply represented.

It is worth mentioning that in principle, using simple neural networks to predict events that have not yet started or finished is feasible since neural networks can automatically extract features. However, individual player information alone is insufficient to accurately predict events. For example, whether a player will shoot depends not only on their distance from the goal but also on the level of defense interference. Similarly, whether a player will pass the ball depends not only on their own position but also on the positions of their teammates and possible interference by opposing players along the passing route. Three essential elements cohesively contribute to the occurrence of an event: the number of objects in the environment, the attributes (location, speed, and posture) of the objects, and the relationships between these objects.

Inspired by the OAR approach [42, 45], we construct the soccer game as a graph with players and balls as objects. In this graph, the information of each object itself is represented as a vector of attributes, while the interactions between different objects are abstracted as relationships. In a study by Li et al. on pedestrian trajectory prediction [26], similar graphs were constructed and analyzed using graph neural networks. These graphs are composed of pedestrian targets because graph neural networks could explicitly use information about the relationships between objects in the graph for inference. Soccer events involve more complex relationships between different objects, requiring careful consideration in designing a graph neural network–based approach.

Based on the analysis of the above factors, we propose a multimodal object-attribute-relation network to address the challenge of soccer event prediction. The network comprises an attribute computation module, a relational attention module, a graph inference module, and an event prediction module. In the attribute computation module, we extract features from visual data and spatiotemporal trajectory data and use these two features as the attribute vectors of the objects. In the relational attention module, we calculate the weights of relationships by considering the relative trajectories between objects. Given the diverse types of relationships between objects, including cooperative relationships between teammates and adversarial relationships between opponents, we employ the multihead GRU attention mechanism to compute different types of relation weights for each object pair. While heterogeneous graphs can handle scenarios with multiple objects and relationships, the multihead attention mechanism offers a more concise approach. The graph reasoning module allows global context information available to each object through message passing, enabling the computation of object embeddings. In the event prediction module, we define single-object and two-object events. The probability of a single-object event is calculated based on the embedding vector of one object, while the probability of a two-object event is calculated based on the embedding vector of a pair of objects. To evaluate the effectiveness of the OAR network, we conduct extensive experiments on the two most representative soccer events, passes and shots, on the SkillCorner and soccER datasets, and our method is able to achieve an accuracy of over 95% on the challenging task of event intention prediction, outperforming traditional methods.

In summary, this work makes three main contributions:

(1) We design an object-attribute-relation graph with multimodal information to represent soccer games. Unlike existing methods that primarily focus on spatiotemporal data without incorporating visual information, our approach utilizes ConvLSTM to extract movement features from video data and combines them with trajectory features to represent object attributes. The integration of multimodal information plays a positive role in the following graph reasoning:
(2) We develop an OAR graph model with a multihead GRU attention mechanism to incorporate heterogeneous and multi-objective information into an end-to-end event prediction framework. Compared to existing schemes, our framework explicitly models the multidimensional interactions between objects, enabling the analysis of complex object intentions.
(3) We evaluate the proposed OAR network on SkillCorner and soccER datasets to demonstrate its effectiveness on real-game data. The extensive experimental results demonstrate that the proposed method achieves the state-of-the-art performance.

3.2.2 *Related Works*

3.2.2.1 Object-Attribute-Relation Model

The OAR model was first proposed by Wang et al. and initially used in cognitive informatics [42]. The OAR model aims to hierarchically model concepts in human cognition to address semantic computing challenges in various domains, including multimedia computing [36], semantic webs [6], computational linguistics [29], software science [53], and denotational mathematics. It represents a graph data structure where relationships exist between objects, between objects and attributes, and between attributes themselves. Such a graph structure provides a rigorous means to characterize multimedia information, such as images [32], text [44], and program code [43]. Inspired by the OAR model, we adopt a similar graph structure to represent the soccer game.

3.2.2.2 Soccer Match Analysis

Soccer match analysis plays a vital role in decision-making processes for coaches, analysts, scouts, and managers. Silvio et al. presented a large and scalable dataset for many soccer-related tasks, such as player detection, action spotting, and replay grounding [8]. Numerous methods based on different data sources have been developed for soccer event recognition. Wickramaratna et al. used neural networks to detect goal events from video data [46]. Richly et al. compared different machine learning approaches to recognize passing and kicking events [34, 35]. Lia et al. used declarative interval temporal logic to formalize soccer events and build a soccer simulation dataset [30]. Huang et al. analyzed semantic information to automatically detect goals, penalties, corner kicks, and card events [15]. Tavassolipour et al. used Bayesian networks to summarize games [39]. While event recognition and game summarization are essential preprocessing steps for tactical analysis, research on player-team habit analysis, match prediction, and soccer AI has proven to be of significant value. Yue et al. used statistical methods to evaluate player and team behavior in a soccer match based on spatiotemporal data [51, 52]. Bialkowski et al. analyzed the strategy on a large dataset for one complete soccer season [3]. Kim et al. presented a system that predicted short-term future ball positions based on motion fields calculated from video [20]. Chen et al. presented a multiagent soccer AI on Google research football [16, 23].

3.2.2.3 Graph Neural Networks

Graph neural networks were first proposed by Gori [9] and were first used to process data with graph structures, such as knowledge graphs and chemical molecules. Initially, graph neural networks learned the representation of target nodes by iteration. Inspired by CNNs, convolutional graph neural networks, which model interactions through information aggregation, have become the mainstream approach. Graph convolutional network (GCN) [22] is the representative method of convolutional graph neural networks. Graph attention network (GAT) [41] adds an attention mechanism to GCN. Shapley additive global importance (SAGE) [13], gated graph neural networks (GGNN) [27], and graph isomorphism networks (GIN) [49] modify the message passing function to achieve better performance. The graph neural networks can be applied to multimedia tasks, such as scene graph generation [47, 50], visual question answering [31], and video summarization [54]. Nicholas, Li et al. applied graph neural networks to handle multiagent interaction problems, such as pedestrian trajectory prediction [12, 25]. Inspired by NRI [21], Li et al. used dynamic relational inference and presented the EvolveGraph [26]. Since soccer events are composed of player behaviors, and player trajectories are analogous to pedestrian trajectories, employing a graph neural network becomes a promising solution for soccer event prediction.

3.2.2.4 Attention Mechanisms

Attention in deep neural networks is widely used for visual tasks and natural language processing. In the image caption task [48], the attention model attempts to align words with different regions of the image based on their similarity. The transformer model [40] was first proposed in natural language processing, which considers some similarity between all words in a sentence and describes different types of similarities using a multihead attention mechanism. The transformer model is also applied in image classification, where vision transformer (VIT) [7] models the relationship between regions of an image and achieves substantial performance gains. Social attention [1, 11] is used in multiagent interaction scenarios. Essentially, attention is the weight of the relationship between two objects, and these methods differ only in the nature of the objects being images, words, or other entities. The soccer game we are studying can be considered a dynamic multiagent interaction scenario, where attention weights between two objects change dynamically. As a result, we use the changing trajectory information to capture the evolving relationships between different objects.

3.2.3 The Proposed Method

We propose a framework called OARNet to address the problem of soccer event prediction. In a soccer game, the behavior of an object is influenced not only by its own state but also by other objects. Additionally, the number of objects involved in an event is uncertain. Therefore, a graph structure is well suited for representing a clip of a soccer game. In this graph, objects such as players are abstracted as nodes. We categorize soccer events into single-player events and two-player events. Single-player events, like shooting, are performed by a single player, while two-player events, such as passing, involve two players. Consequently, single-player events can be represented as attributes of the graph nodes, and two-player events can be represented as relations in the graph. The task of event prediction is then transformed into node classification and relation classification within the graph.

To construct the graph, first, spatiotemporal data is fed into a long short-term memory networks (LSTM) [4, 14] to extract spatiotemporal representations. Subsequently, video data is fed into a convolutional long short-term memory networks (ConvLSTM) [37] to extract video representations. These representations make up the initial attributes of the objects. Next, we propose the multihead GRU attention module to calculate the weights of relations. Then, a graph neural network (GNN) is used to reason out the contextual information among objects. Finally, new object representations updated by the GNN are processed by an event predictor network to generate event predictions. Figure 3.3 elaborates the overview of the OARNet. The data format and the method for computing multimodal representations will be introduced in Sect. 3.2.3.1, followed by detailing of multihead GRU attention and graph reasoning methods in Sect. 3.2.3.2. Finally, the optimization objective of the framework is introduced in Sect. 3.2.3.3.

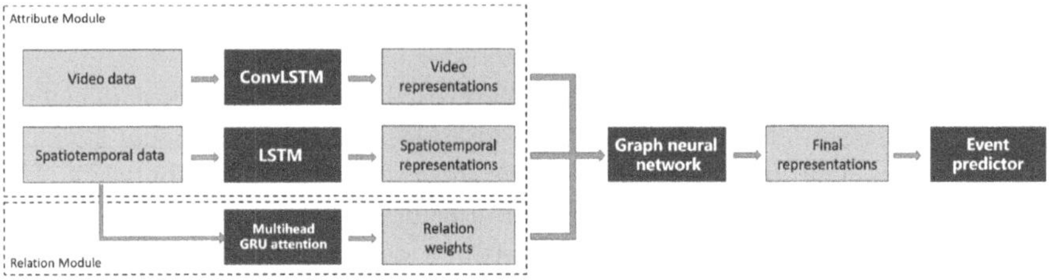

Fig. 3.3 An overview of the proposed OARNet. In the attribute module, video representations and spatiotemporal representations are extracted by ConvLSTM and LSTM, respectively, and they are concatenated to form the initial attribute of OAR graph. Relation weights are calculated from spatiotemporal data by multihead GRU attention. The graph neural network produces the final representations of objects. The final representations are fed into an event predictor network to predict the possibilities of single-player events and two-player events. See texts for more details

3.2.3.1 Multimodal Representation

To enable the subsequent graph computation to better capture event-related information, we incorporate two modalities: spatiotemporal data and video data. While previous methods primarily focus on spatiotemporal data, which only provides location information, video data contain valuable information about the pose of objects, which is significant for analyzing object state. We denote the spatiotemporal information of the ith object at the tth frame as $x_{i,t}^{sp}$, which is a vector describing the horizontal and vertical coordinates, the team to which the object belongs, and other optional states. Thus, the spatiotemporal data of a clip with T_j frames and N_o objects can be represented as $\{\{x_{1,1}, x_{1,2}, \ldots, x_{1,T_f}\}, \{x_{2,1}, x_{2,2}, \ldots, x_{2,T_f}\}, \ldots, \{x_{N_o,1}, x_{N_o,2}, \ldots, x_{N_o,T_f}\}\}$. Similarly, we use I_i, to denote the cropping of the video at the tth frame which corresponds to the i-th object. The video data can be represented as $\{\{I_{1,1}^{\text{video}}, I_{1,2}^{\text{video}}, \ldots, I_{1,T_f}^{\text{video}}\}, \ldots, \{I_{N_o,1}^{\text{video}}, I_{N_o,2}^{\text{video}}, \ldots, I_{N_o,T_f}^{\text{video}}\}\}$.

We define an OAR graph to describe a clip of a soccer game. The OAR graph is denoted as $\mathcal{G}_{OAR} = (\mathcal{V}_{OAR}, \mathcal{E}_{OAR})$, where $\mathcal{V}_{OAR}$ is a set of $N_p + 1 + 1$ objects corresponding to N_p players and 1 ball and v_i is the ith object. The relation between the mth and nth nodes, denoted as $e_{m,n} \in \mathcal{E}_{OAR}$, represents the weights of the relation. $e_{m,n} \in \mathbb{R}^q$ is a vector that captures the probability of different kinds of relations, considering that relations can be multimodal. It is unnecessary and redundant to add every possible relation in the OAR graph. For example, relations between goalkeepers are not meaningful. We use a player-centered approach to select initial relations in the graph. For each player, relations between the player and the Nk nearest players are added to the graph. Besides, relations between the ball and all players are added. The distance between two players is defined as the average Euclidean distance over the course of the clip.

An initial representation consists of a spatiotemporal representation and a video representation. The spatiotemporal representation is calculated by a three-layer LSTM. For the i-th object, the LSTM takes the spatiotemporal data

$\{x_{i,1}, x_{i,2}, \ldots, x_{i,T_f}\}$ of T_f frames as input and produces three sets of hidden states $\{h_{i,t,1}\}, \{h_{i,t,2}\}, \{h_{i,t,3}\}$. The spatiotemporal representation vector $f_i^s \in \mathbb{R}^{d_{out}^s}$ is obtained by concatenating the three hidden states of the last frame:

$$\begin{aligned} h_{i,t,1} &= \text{LSTM}_1(x_{i,t}, h_{i,t-1,1}) \\ h_{i,t,2} &= \text{LSTM}_2(h_{i,t,1}, h_{i,t-1,2}) \\ h_{i,t,3} &= \text{LSTM}_3(h_{i,t,2}, h_{i,t-1,3}) \\ f_i^s &= \text{Concat}(h_{i,T_f,1}, h_{i,T_f,2}, h_{i,T_f,3}) \end{aligned} \tag{3.1}$$

For the ith object, the ConvLSTM takes $\{I_{i,1}^{\text{video}}, I_{i,2}^{\text{video}}, \ldots, I_{i,T_f}^{\text{video}}\}$ as input and produces T_f hidden feature maps $\{M_{i,t}\}$. We apply max pooling to the last hidden feature map and flatten it to obtain the video representation vector as follows:

$$\begin{aligned} M_{i,t} &= \text{ConvLSTM}\left(I_{i,t}^{\text{video}}, M_{i,t-1}\right) \\ f_i^v &= \text{Max Pooling}(M_{i,T_f}) \end{aligned} \tag{3.2}$$

Finally, the spatiotemporal representation vectors and the video representation vectors are concatenated to form the initial attributes of the objects in the OAR graph. The multimodal representation module is shown in Fig. 3.4.

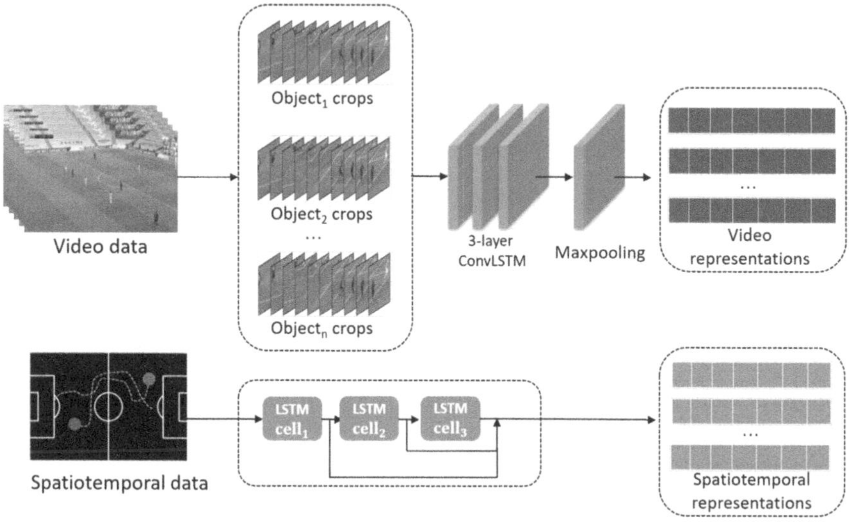

Fig. 3.4 Structure of multimodal representation module

3.2.3.2 Graph Reasoning

We employ a modified two-layer SAGE network [13] to update the representations of each object. The message passing operation in an SAGE layer is performed as follows:

$$X_l^{ob} = \sigma\left(\left[E_l * X_{l-1}^{ob}; X_{l-1}^{ob}\right] W_l + b_l\right) \tag{3.3}$$

Here, $X_l^{ob} \in \mathbb{R}^{n \times d_l}$ is the hidden features of n objects in the lth SAGE layer, where d_l is the dimension of the hidden feature. $X_0^{ob} \in \mathbb{R}^{n \times d_0}$ is initialized as $[f_1^0; f_2^0; \ldots; f_n^0]$, and f_i^0 is acquired by $f_i^0 = Concat(f_i^s, f_i^v)$. $W_l \in \mathbb{R}^{2d_{l-1} \times d_l}$ and $b_l \in \mathbb{R}^{d_l}$ are the weight matrix and the bias vector of a linear layer, respectively. For each layer, $\sigma(\cdot)$ is a nonlinear activation function elu [5]. $E_l \in \mathbb{R}^{n \times n}$ is the adjacent matrix associated with G. The element $e_{i,j}$ of E_l is the weight of relation between the i-th object and the j-th object. The final updated representation matrix is $X_l^{ob} = [f_1^l; f_2^l; \ldots f_n^l]$, where N_l is the number of layers in the graph reasoning network.

We propose GRU attention to calculate the weights of relations. The GRU takes the spatiotemporal data of the two objects associated with a relation as input. It is noteworthy that using pairwise similarities of object representations might not always be suitable for representing the weights of relations. For example, there may exist a strong relation between two players involved in a long pass event, even if their situations and states are quite different. Besides, the information of the ball is also important, so the trajectories of the ball are fed into the GRU. GRU is a simplified version of LSTM, and both can be used to process trajectory features in sequential form. We compared the performance of GRU and LSTM in preliminary experiments, and the difference was not significant. Considering that the attention module will be used several times in our model, we opt for GRU instead of LSTM to reduce the model parameters. The weight of the relation between the i-th object (source object) and the j-th object (destination object) is calculated as follows:

$$\begin{aligned} h_i &= GRU\left(\left[x_{i,t}^{sp}; x_{b,t}\right]\right) \\ e_{i,j} &= Sigmoid([h_i; h_j]W_e + b_e) \end{aligned} \tag{3.4}$$

Here, W_e and b_e represent the weight matrix and the bias vector of a linear layer, respectively, and $x_{b,t}$,t represents the trajectories of the ball.

Considering that relations are multimodal, we apply the multihead mechanism to GRU attention. The multihead GRU attention is depicted in Fig. 3.5. With multihead GRU attention, the message passing in the lth layer is modified as follows:

$$\begin{aligned} head_{l,i} &= \sigma([E_{l,i} * X_{l-1}^{ob}; X_{l-1}^{ob}]W_{l,i} + b_{l,i}) \\ X_l^{ob} &= \text{Concat}(head_1, head_2, \ldots, head_h) \end{aligned} \tag{3.5}$$

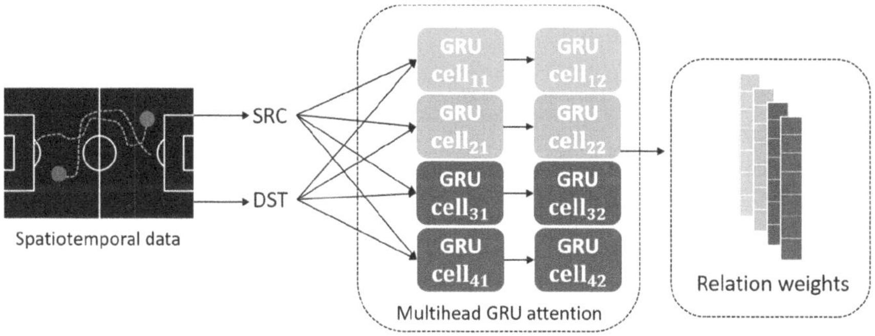

Fig. 3.5 Structure of multihead GRU attention

The introduction of the multihead attention mechanism in multimodal prediction tasks for soccer events offers a particular advantage by efficiently and concisely handling heterogeneous graphs. Different OAR graphs can be constructed based on different prior knowledge. For example, a graph containing l nearest neighbors of each object can be constructed to model the influence of objects surrounding each player. A complete graph containing only players from the same team can be constructed to represent support and collaborative relationships. Similarly, a complete graph containing only players near the ball can be considered to contain the most critical objects. Different approaches to OAR graph construction reflect heterogeneity. However, there is coupling and redundancy between these graphs, where relations in different graphs may be of the same type and relations in the same graph may be of different types. Traditional heterogeneous graph networks are not suitable for processing multiple types of OAR graphs since they lack an explicit type of relation. By employing the multihead attention mechanism, we can easily address this issue by using different OAR graphs for different heads.

Using the graph neural network, the representation of each object is updated through message passing from the surrounding objects. Subsequently, we employ multilayer perceptrons (MLPs) with a sigmoid activation function to predict events. For single-player events involving the i-th object, the MLP directly takes f_i^l as input and produces $\hat{y}_i \in [0, 1]$ as the predicted probability value of the event. For two-player events from the i-th object to the j-th object, the MLP takes f_i^l and f_j^l as inputs and produces $\hat{y}_{i,j} \in [0, 1]$. The probabilities of single-player event $\hat{y}_i$ and two-player event $\hat{y}_{i,j}$ are calculated as follows:

$$
\begin{aligned}
\hat{y}_i &= \text{Sigmoid}\left(\text{MLP}\left(f_i^l\right)\right) \\
\hat{y}_{i,j} &= \text{Sigmoid}\left(\text{MLP}\left(f_i^l; f_j^l\right)\right)
\end{aligned}
\tag{3.6}
$$

The structure of one graph reasoning layer with multihead GRU attention is illustrated in Fig. 3.6.

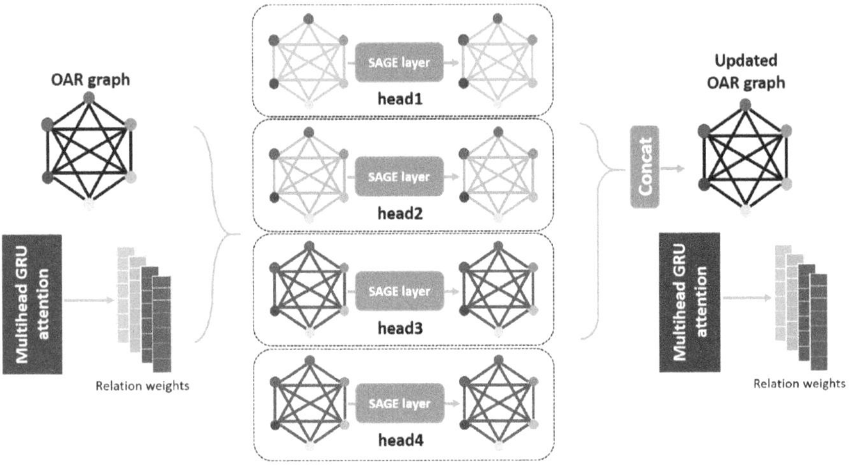

Fig. 3.6 Structure of one graph reasoning layer with multihead GRU attention. In this layer, the number of heads is 4

3.2.3.3 Optimization Objective

In each event, a player makes a decision only once, but there are often multiple options available. For example, when the ball carrier has multiple suitable teammates to pass to, they can only choose one. This means that there is a certain possibility of alternative outcomes. To enhance the generalization of our model, we utilize a loss based on binary cross entropy (BCE) loss:

$$\mathcal{L}_\alpha(y, \hat{y}) = -y\log\hat{y} + (1-y)\hat{y}^\alpha \tag{3.7}$$

where $y = 0, 1$ is the ground-truth and $\hat{y}$ is the predicted probability. When the event happens, $y = 1$ and $\mathcal{L}_\alpha(y, \hat{y}) = -log\hat{y}$ is BCE loss, otherwise $y = 0$ and $\mathcal{L}_\alpha(y, \hat{y}) = \hat{y}^\alpha$. The power function limits the range of values between [0, 1] and prevents it from converging to infinity as $\hat{y}$ approachs to 1 compared to the original $-log(1-\hat{y})$.

In single-player events like shots and two-player events like passes, the player in possession of the ball is the most concerned object. The probabilities of shot and pass events which are not executed by the ball carrier v_c are definitely 0. However, there is no guarantee that the data indicate which object is v_c. So we make a further adjustment to the loss. For single-player events, the loss is formulated as follows:

$$\mathcal{L}_{sp} = \mathcal{L}_\alpha(y_c, \hat{y_c}) - \lambda_1 \sum_{i \neq c} \log(1-\hat{y}_i) \tag{3.8}$$

For two-player events, let $\mathcal{R}_p$ be the set of relations where the source node is v_c and both the source and destination nodes are in the same team. The loss is then formulated as:

$$\mathcal{L}_{tp}^{\text{event}} = \sum_{r_{i,j} \in \mathcal{R}_p} \mathcal{L}_\alpha(y_{i,j}, \hat{y_{i,j}}) - \lambda_2 \sum_{r_{i,j} \notin \mathcal{R}_p} \log(1 - \hat{y}_{i,j}) \tag{3.9}$$

3.2.4 Experiments and Analyses

3.2.4.1 Datasets

In this study, we evaluate our object-attribute-relation-based soccer event prediction methods on two datasets: the SoccER dataset and the SkillCorner dataset.

(1) *soccER Dataset*: This dataset was proposed by Morra et al. [30]. It is a synthetic dataset built from the Gameplay Football Engine [23], which can simulate a complete soccer game and generate both videos and trajectories of objects in the game. Although the graphics are not photorealistic, the player actions and models are relatively realistic. Player movement and behavior are generated in one of three modalities: human player vs. human player, human player vs. AI, or AI vs. AI. The trajectories and bounding boxes of objects are highly accurate, which benefits the stability of our method.

The trajectory data are provided as a list, where each element corresponds to the tracking result for a frame. Each frame contains up to three types of objects: players, referees, and balls. The dataset provides the 2D coordinates of the objects, as well as their shirt numbers.

The trajectories have a frame rate of 100 fps, while the videos are recorded at 25 fps. To reduce computation, we downsample the trajectories to 25 fps and the videos to 5 fps.

The SoccER dataset consists of 16 games. Among the various events, there are 5059 passes and 506 shots. We use 12 games for training and 4 games for testing.

(2) *SkillCorner Dataset*: This dataset is collected by SkillCorner, a Paris-based start-up specializing in AI tracking. It includes tracking data from nine matches, featuring the champions and runners-up of the 2019/2020 league matches in the English Premier League, French L1, Spanish LaLiga, Italian Serie A, and German Bundesliga. Table 3.1 provides information about the nine games.

The data were automatically collected from the broadcast feed without any manual correction. Therefore, the data are limited to what is visible in the broadcast video, and not all players are visible at all times. The format of the tracking data in the SkillCorner dataset is similar to that of the SoccER dataset. In addition to the shirt numbers and 2D coordinates, some frames indicate which player possesses the ball.

The SkillCorner dataset is based on real games. In reality, spatiotemporal data can be obtained using position sensors or object detection, and video data can be obtained from cameras. The coordinates in the spatiotemporal data are relative to the real court, and the player positions in the video frames are transformed from

Table 3.1 Games information

Home team	Away team	Date	Score
Manchester City	Liverpool	2020-07-02	4-0
Dortmund	Bayern Munchen	2020-05-26	0-1
Juventus	Inter	2020-03-08	2-0
Real Madrid	FC Barcelona	2020-03-01	2-0
FC Barcelona	Real Madrid	2019-12-18	0-0
Liverpool	Manchester City	2019-11-10	3-1
Bayern Munchen	Dortmund	2019-11-09	4-0
Paris	Marseille	2019-10-27	4-0
Inter	Juventus	2019-10-06	1-2

Fig. 3.7 An example of data processing pipeline. (**a**) depicts the input picture, (**b**) depicts the result of court homography estimation, and (**c**) shows the bounding boxes generated based on the spatiotemporal data and the estimated homography matrix

the real coordinates through homography transformation. Let $(x_s, y_s, 1)$ denote the homogeneous coordinates of an object on the court (assuming the object is approximately on the plane of the court and ignoring the z-axis coordinate for simplicity), and let $(x_s', y_s', 1)$ denote the homogeneous coordinates on the image. The homography matrix is denoted as H_s. The transformed coordinates $(x_s', y_s', 1)$ can be calculated as follows:

$$(x_s, y_s, 1) = H_s \left(x_s', y_s', 1\right) \tag{3.10}$$

To process the SkillCorner dataset and real data, the spatiotemporal data and video data need to be aligned. We use the methods proposed by Wei et al. [17] to estimate the homography matrix. With the estimated homography matrix and the coordinates relative to the court, we can locate the objects in the image. For simplicity, we assume the bounding box of each object has the same size, and set the transformed coordinate $(x_s', y_s', 1)$ as the midpoint of the lower edge of the bounding box. An example of the data processing pipeline is shown in Fig. 3.7.

The frame rate of the broadcast video is 25 fps. To synchronize the video data with the spatiotemporal data, we downsample both to 5 fps. We use seven games for training and two games (2020-07-02 and 2020-05-26) for testing. Passing events and shooting events are manually labeled for these games, resulting in 937 passing events and 119 shooting events in the training set, and 311 passing events and 39 shooting events in the test set.

3.2.4.2 Experimental Setup

We compare the performance of our OAR method with the following baseline models: (1) Support vector machine (SVM), which extracts features such as position, velocity, and distance to the ball from spatiotemporal data and uses an SVM for classification; (2) MLP, which directly classifies sampled trajectories; (3) LSTM+MLP, which utilizes LSTM to extract features from trajectories and MLP for player and ball feature classification and further event prediction; (4) object detection, which detects events using rules based on temporal and logical combinations of detected objects and their relative distances; (5) temporal logic, which employs a two-tier declarative interval temporal logic system to detect atomic and complex events.

Implementation Details In the default configuration, we use a player-centered method to select initial edges. Both the LSTM for spatiotemporal representation and the ConvLSTM for video representation have three layers. The default graph reasoning network consists of a two-layer SAGE network with multihead GRU attention. The MLP for final event prediction has two layers. In each sub-experiment, only one module is replaced, while the other parts remain consistent with the default configuration.

We evaluate five tasks: pass recognition, pass prediction, pass intention prediction, shot recognition, and shot intention prediction. Pass recognition refers to identifying the passer and receiver when a pass is completed and the ball is already under the control of the receiver. At this point, the model can gather all the necessary information of the event and determine whether it is a passing event or not. Pass prediction is for the scenario when the pass is in progress and the ball is between the passer and the receiver. In pass prediction task, the model needs to predict the target object of the pass based on the ball movement and positions of the surrounding players. Pass intention prediction is for the scenario when the pass has not occurred yet and the ball is still at the passer's feet and the model needs to predict whether a pass will happen and which is the most suitable target for the pass based on the player's position distribution. These three tasks all occur during the passing event, with the difference lying in the stage of the passing process being performed. Shot recognition involves identifying the player who shoots when the shooting action is finished, while shot intention prediction predicts whether a player intends to shoot when the ball is still at the player's feet. For passing events, the ratio of positive samples, samples where the receiver is wrong, and samples where no passing occurs is 1:1:1. For shooting events, the ratio of positive samples, samples where the shooter is incorrect, and samples where no shooting occurs is 1:1:2.

Four metrics, including accuracy, precision, recall, and F1-score, are adopted for evaluation.

Table 3.2 Performance comparison on the soccER dataset

Task	Method	Acc.	F1.	Prec.	Rec.
Pass recognition	SVM	48.8%	36.2%	31.0%	43.6%
	MLP	89.9%	86.6%	77.2%	98.7%
	LSTM+MLP	66.4%	59.4%	49.7%	73.9%
	Object detection	–	89%	94%	84%
	Temporal logic	–	94%	96%	93%
	OAR method	**99.2%**	**98.7%**	**99.8%**	**97.6%**
Shot recognition	SVM	49.2%	33.7%	25.0%	51.6%
	MLP	93.8%	87.3%	90.0%	84.7%
	LSTM+MLP	90.8%	82.8%	77.8%	88.4%
	Temporal logic	–	80%	74%	89%
	OAR method	**96.4%**	**93.3%**	**89.2%**	**97.9%**
Pass prediction	SVM	55.5%	45.2%	38.4%	54.9%
	MLP	71.8%	64.0%	55.7%	75.1%
	LSTM+MLP	75.2%	59.4%	49.7%	73.9%
	OAR method	**96.5%**	**94.9%**	**91.6%**	**98.5%**
Pass intention prediction	SVM	53.3%	34.1%	32.2%	36.3%
	MLP	69.2%	61.0%	52.7%	72.5%
	LSTM+MLP	69.0%	64.7%	52.1%	85.5%
	OAR method	**87.3%**	**83.7%**	**75.2%**	**94.4%**
Shot intention prediction	SVM	52.1%	33.9%	48.8%	26.0%
	MLP	89.0%	82.0%	70.1%	**98.8%**
	LSTM+MLP	93.8%	88.6%	82.9%	95.1%
	OAR method	**95.7%**	**92.3%**	**87.8%**	97.3%

Bold values means the best metric in the comparison algorithms

$$\begin{aligned} \text{Precision} &= \frac{\text{TP}}{\text{TP}+\text{FP}}, \text{Recall} = \frac{\text{TP}}{\text{TP}+\text{FN}} \\ \text{Accuracy} &= \frac{\text{TP}+\text{TN}}{\text{TP}+\text{FP}+\text{TN}+\text{FN}} \\ \text{F1}-\text{score} &= 2 \times \frac{\text{Precision} \times \text{Recall}}{\text{Precision}+\text{Recall}} \end{aligned} \tag{3.11}$$

where TP, FP, TN, and FN represent true positive, false positive, true negative, and false negative cases, respectively.

3.2.4.3 Effectiveness of Involving Object-Attribute-Relation Graphs in Soccer Event Recognition and Prediction

Table 3.2 shows the results of different methods on the SoccER dataset. According to [30] and [34], the SVM method can identify some passing events with a

Table 3.3 Performance comparison on the SkillCorner dataset

Task	Method	Acc.	F1.	Prec.	Rec.
Pass prediction	MLP	48.0%	47.3%	35.7%	70.0%
	LSTM+MLP	57.0%	26.0%	30.5%	22.7%
	OAR method	**83.8%**	**85.9%**	**80.5%**	**92.1%**
Shot recognition	MLP	**85.8%**	4.9%	19.2%	2.8%
	OAR method	80.1%	**89.7%**	**89.5%**	**90.0%**
Pass intention prediction	MLP	56.8%	39.8%	37.1%	42.9%
	LSTM+MLP	54.1%	38.1%	34.6%	42.4%
	OAR method	**71.2%**	**84.8%**	**78.9%**	**91.7%**
Shot intention prediction	MLP	**85.6%**	0.0%	0.0%	0.0%
	OAR method	74.9%	**87.3%**	**80.1%**	**96.0%**

Bold values means the best metric in the comparison algorithms

precision of 42.6% and an F1-score of 51%. However, the performance of the SVM method on the SoccER dataset significantly deteriorates, indicating its limited ability in event recognition. The results in Richly et al. [34] heavily rely on a well-designed feature extraction method. Neural networks, such as MLP and LSTM, can automatically extract features, and their performance proves that they learn at least some simple patterns that contribute to event recognition. The F1-score of MLP reaches 86.6% in pass recognition and 87.3% in shot recognition. By incorporating visual information, the object-detection-based method achieves an F1-score of 89% in pass recognition. The temporal logic method performs relatively well. Our OAR method shows significant improvement in pass recognition, and its performance in shot recognition is also acceptable. Event recognition is not a very challenging task, so most methods achieve fairly good results.

For event prediction, the SVM method still fails to learn useful information. Object-detection-based and temporal logic methods are not designed for prediction tasks as they rely on explicit movement and action characteristics. Therefore, we compare neural network methods. In pass prediction and pass intention prediction, the OAR method improves accuracy by 20.0% and 11.3%, respectively, over the LSTM+MLP method. This demonstrates the necessity of involving graph reasoning since prediction tasks heavily depend on global context information, as each object is influenced by other objects. In shot intention prediction, the OAR method maintains high accuracy, although its performance is not significantly better than that of the LSTM+MLP method. We speculate that this is because a player's intention to shoot is more influenced by individual factors, and the difference between most negative samples and positive samples is trivial because holding the ball at a certain distance from the goal is almost equivalent to shooting, and challenging samples are rare.

Due to the presence of noise in SkillCorner data and the instability of the data alignment method, the overall performance of all methods decreases, as shown in Table 3.3. MLP and LSTM+MLP struggle to converge, particularly because shot events are scarce. Overall, the OAR method outperforms the other methods.

Table 3.4 Performance of using different visual information processing method

	PR		PP		PIP		SR		SIP	
Method	acc.	F1.	acc.	F1.	acc.	F1.	acc.	F1.	acc.	F1.
Without video	97.7%	96.3%	94.4%	92.6%	79.8%	74.0%	**93.9%**	87.8%	92.3%	87.8%
VGG feature of last frame	97.8%	96.7%	93.4%	92.8%	79.9%	72.6%	92.3%	**88.2%**	90.9%	85.1%
Posture feature	97.5%	96.9%	92.3%	91.1%	82.7%	75.6%	93.0%	86.3%	91.2%	86.4%
ConvLSTM	**98.1%**	**97.2%**	**96.0%**	**94.2%**	**85.9%**	**80.3%**	92.7%	86.2%	**94.9%**	**90.0%**

Bold values means the best metric in the comparison algorithms

3.2.4.4 Contributions of Visual Information

We compare four methods of visual information processing. (1) We remove the visual processing branch. (2) We use the pose recognized by OpenPose [33] as visual features. (3) We extract visual features using the feature layers of VGG [38] pretrained on ImageNet. Since VGG features are high-dimensional and abstract, we only perform feature extraction on the last frame. (4) We use ConvLSTM as the visual feature extraction module.

Table 3.4 presents the performance of different visual information processing methods. Compared to the model without visual information, using ConvLSTM shows some improvement, while using VGG features and pose features does not exhibit a significant improvement. The VGG features are too abstract and do not accurately describe details or incorporate dynamic information. Additionally, the image resolution may not be high enough, and the presence of overlapping individuals leads to non-negligible noise in the pose obtained from OpenPose. The simple pose features also overlook the importance of ball information, which is crucial for judging ball holding and kicking actions. ConvLSTM, when combined with other modules in an end-to-end system, allows for joint optimization of all modules. As a relatively lightweight model suitable for processing sequence image data, ConvLSTM compensates for the lack of important action features in trajectory information.

3.2.4.5 Effectiveness of Attention Mechanisms

The attention mechanism is critical for OARNet. Each object in the object-attribute-relation graph is influenced by other objects to varying degrees, and the relation weights calculated from the attention mechanism to some extent describe these degrees. To investigate the effectiveness of different attention mechanisms, we evaluate six methods in our experiment.

(1) Average attention: In this experimental setup, all relation weights are set to 1, which is equivalent to GCN without normalization.
(2) Random attention: In this experimental setup, all relation weights are set to random values in the range of [0, 1].

Table 3.5 Performance of using different attention mechanisms

	PR		PP		PIP		SR		SIP	
Method	acc.	F1.	acc.	F1.	acc.	F1.	acc.	F1.	acc.	F1.
Avg (GCN)	93.3%	90.7%	86.3%	82.7%	76.3%	70.0%	91.3%	84.3%	92.0%	86.0%
Random	90.6%	87.1%	85.6%	81.6%	74.9%	68.8%	93.4%	87.7%	92.2%	86.1%
GAT	**98.7%**	**98.1%**	95.0%	92.9%	79.4%	74.4%	**94.0%**	87.7%	93.5%	88.4%
GRU attention	98.1%	97.2%	95.2%	93.2%	80.3%	74.4%	**94.0%**	**87.9%**	93.1%	87.9%
Multihead GAT	97.9%	97.0%	95.0%	92.9%	80.9%	74.5%	91.5%	84.6%	93.2%	88.2%
Multihead GRU attention	98.1%	97.2%	**96.0%**	**94.2%**	**85.9%**	**80.3%**	92.7%	86.2%	**94.9%**	**90.0%**

Bold values means the best metric in the comparison algorithms

(3) GAT: For simplicity, we modify the attention formula of GAT, and the relation weights are calculated as follows:

$$e_{i,j} = \text{Sigmoid}(W * [f_i; f_j] + b) \tag{3.12}$$

where W and b are the weight matrix and the bias vector of a linear layer, respectively. f_i represents the hidden feature of the ith object.
(4) GRU attention: The number of heads in multihead GRU attention is set to 1, and the dimension of the graph layer is increased by four times to make the number of parameters almost equal to that of the multihead GRU attention model.
(5) Multihead GAT: This method applies a multihead mechanism to GAT, with the number of attention heads set to 4.
(6) Multihead GRU attention: This is the attention mechanism proposed with the number of attention heads set to 4.

Table 3.5 shows the effectiveness of attention mechanisms. The performances of average and random attention are significantly worse than those of any other attention method. The difference between GAT and GRU attention is not substantial. Applying the multihead mechanism to GAT leads to a slight decline in accuracy and F1-score in pass recognition, shot recognition, and shot intention prediction. This can be attributed to the multihead attention module having more parameters than the ordinary attention module, which makes it prone to underfitting with the same number of training epochs. However, applying the multihead mechanism to GRU attention improves the performance in pass prediction, pass intention prediction, and shot intention prediction. This indicates that GRU attention is a reasonable design choice and can adapt well to the multihead mechanism.

3.2.4.6 Impact of Message Passing Functions

Different message passing functions also influence the performance of the graph model. In this section, we explore five message-passing functions obtained from the Deep Graph Library (DGL).

(1) GraphConv

$$X_n^{ob} = \sigma\left(X_{n-1}^{ob} W_n + b_n\right) \tag{3.13}$$

(2) GATConv

$$X_n^{ob} = E\sigma\left(X_{n-1}^{ob} W_n + b_n\right) \tag{3.14}$$

(3) SAGEConv

$$X_n^{ob} = \sigma\left(\left[EX_{n-1}^{ob}; X_{n-1}^{ob}\right] W_n + b_n\right) \tag{3.15}$$

(4) GINConv

$$X_n^{ob} = \sigma\left(\left((1+\epsilon)EX_{n-1}^{ob} + X_{n-1}^{ob}\right) W_n + b_n\right) \tag{3.16}$$

where ϵ is a learnable parameter.

(5) GatedGraphConv

$$X_n^{ob} = \mathrm{GRU}\left(EX_{n-1}^{ob}, X_{n-1}^{ob}\right) \tag{3.17}$$

Table 3.6 presents the performance of all message passing functions on the soccER dataset. In pass prediction and pass intention prediction, the accuracy and F1-score of SAGEConv are considerably higher than those of other functions. This suggests that features of the objects themselves should be retained. Although using a linear layer or information aggregation as the first step is essentially the same,

Table 3.6 Performance of using different message passing functions

	PR		PP		PIP		SR		SIP	
Method	acc.	F1.	acc.	F1.	acc.	F1.	acc.	F1.	acc.	F1.
GATConv	97.9%	97.0%	95.0%	92.9%	80.9%	74.5%	91.5%	84.6%	93.2%	88.2%
GraphConv	93.3%	90.7%	86.3%	82.7%	76.3%	70.0%	91.3%	84.3%	92.0%	86.0%
GatedGraphConv	96.3%	94.7%	92.6%	89.9%	80.4%	74.8%	91.9%	85.4%	93.2%	88.4%
GINConv	**98.1%**	**97.2%**	91.0%	87.7%	77.5%	71.4%	**93.8%**	**87.9%**	92.7%	87.6%
SAGEConv	**98.1%**	**97.2%**	**96.0%**	**94.2%**	**85.9%**	**80.3%**	92.7%	86.2%	**94.9%**	**90.0%**

Bold values means the best metric in the comparison algorithms

Table 3.7 Performance under different data fragment lengths

	0.1s(1 frame)		0.5 s		1.0 s		1.5 s		2.0 s	
Task	acc.	F1.	acc.	F1.	acc.	F1.	acc.	F1.	acc.	F1.
Pass recognition	93.0%	90.2%	97.7%	96.6%	98.1%	97.2%	99.2%	98.7%	95.8%	94.1%
Shot recognition	84.7%	75.9%	87.7%	75.6%	93.4%	88.5%	96.4%	93.3%	92.5%	86.1%
Pass prediction	90.3%	86.6%	95.9%	94.2%	98.5%	97.8%	96.5%	94.9%	97.4%	96.2%
Pass intention prediction	82.2%	76.5%	82.2%	77.3%	83.1%	78.9%	87.3%	83.7%	85.8%	81.6%
Shot intention prediction	86.7%	79.4%	91.2%	85.1%	92.7%	86.9%	95.7%	92.3%	95.2%	91.2%

the latter performs better when the number of graph layers is smaller. This is because the initial features of the first layer are already processed by the multimodal representation module, and there is no immediate need to send them to the linear layer. Hence, the linear layer of the first graph layer in GraphConv and GATConv is less effective than that of SAGEConv and GINConv. SAGEConv concatenates features of the objects themselves with features of their neighbors, while GINConv adds them. The results indicate that the former is superior.

3.2.4.7 Impact of the Length of the Data Fragment

The input of OARNet contains data from two modalities including spatiotemporal data and video data, which are both serialized. The performance of the OARNet model can be affected by the number of frames included in these data. In this section, we evaluate the model's performance using data fragment lengths of 0.1 s, 0.5 s, 1 s, 1.5 s, and 2 s, respectively. The data fragment length of 0.1s contains only one frame. Table 3.7 presents the results.

Our model achieves better results with data fragment lengths of 1.0 s and 1.5 s, while longer data fragments do not necessarily yield the best results. Analyzing the original videos, we find that most passing events are short passes, with the entire process typically taking between 1.0 s and 1.5 s. The time to complete a shooting action is even shorter, usually around 0.5 s. Therefore, we can assume that data fragments of this length already contain the most critical information. Longer data fragments introduce unnecessary information and interfere with the model's judgment. It is worth noting that even with only one frame of data, our model can make relatively reasonable predictions. This is mainly because passes and shots are strongly correlated with static features. For example, a significant proportion of passing events in the dataset occur between the ball carrier and the three closest teammates. The direction of the ball carrier can further exclude unreasonable passing choices. Shot events usually occur in areas without close defenders and at moderate distances. These static features provide sufficient information for prediction in simple situations. Based on the results of this part of the experiment, we choose a data fragment length of 1.5 s as the general experimental setup.

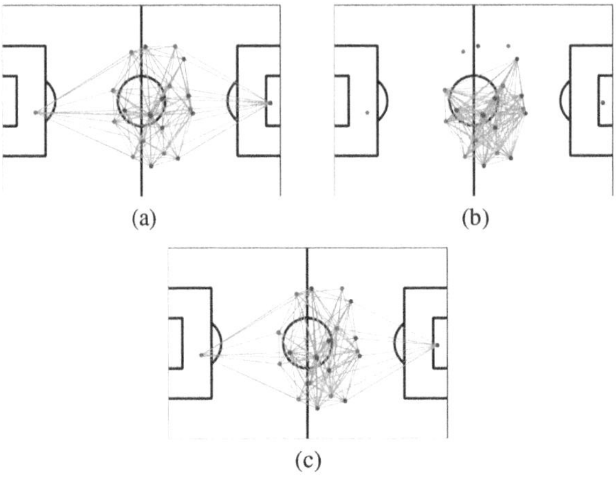

Fig. 3.8 Different initial edges. (**a**) Player-centered. (**b**) Ball-centered. (**c**) Fixed

Table 3.8 Performance of using different initial edges selection methods

	PR		PP		PIP		SR		SIP	
Method	acc.	F1.	acc.	F1.	acc.	F1.	acc.	F1.	acc.	F1.
Player-centered	98.1%	97.2%	96.0%	94.2%	85.9%	80.3%	92.7%	86.2%	94.9%	90.0%
Ball-centered	97.9%	97.0%	79.2%	75.9%	76.0%	72.1%	83.7%	75.1%	89.3%	81.7%
Fixed	87.8%	83.3%	81.1%	75.0%	68.8%	65.0%	93.2%	87.7%	93.2%	87.9%
Mixed	**99.2%**	**98.7%**	**96.5%**	**94.9%**	**87.3%**	**83.7%**	**96.4%**	**93.3%**	**95.7%**	**92.3%**

Bold values means the best metric in the comparison algorithms

3.2.4.8 Impact of Initial Edge Selection Methods

The construction of the object-attribute-relation graph requires the selection of initial edges. We test four edge selection methods: player-centered, ball-centered, fixed, and mixed methods. In the player-centered method, edges between players and their nearest l neighbors at the current frame are selected, where $l = 7$ in this experiment. In the ball-centered method, the players closest to the ball form a complete graph, and the edges in this graph are selected. In the fixed method, edges exist between specific player pairs, such as the forward and the opposing defender, or the forward and their own midfielder. In the mixed method, different graphs are used at each of the four heads in the multihead GRU attention. The four graphs consist of a player-centered graph ($l = 5$), a player-centered graph ($l = 7$), a player-centered graph ($l = 9$), and a merged graph of a player-centered graph ($l = 7$) and a ball-centered graph. The first three methods are illustrated in Fig. 3.8.

Table 3.8 shows the performance of different initial edge selection methods. The player-centered method outperforms the ball-centered and fixed methods. The reason for this is that the ball-centered method fails to capture information about players far from the ball, which is crucial for long passes, and the fixed method is not suitable for rapidly changing objects. The mixed method achieves the best performance as it effectively incorporates information from different OAR graphs.

(a) (b)

(c)

Fig. 3.9 Examples of a pass and a shot on the SkillCorner dataset. (**a**) and (**b**) describe the different phases of a passing event and (**c**) describes a shot that has been completed. Players framed by red bounding boxes are possible passers and players framed by blue bounding boxes are possible receivers. The player framed by a white box is a shooter. The number on the left side of the box is the event probability given by the pass recognition model and the number on the right side is the probability given by the pass prediction model. The number above a white box is the possibility given by the shot recognition model. (**a**) pass prediction. (**b**) pass recognition. (**c**) shot recognition

The player-centered graph focuses on local relations with a small l, while the graph with a large l covers more distant relations. These results indicate that the multihead attention mechanism is suitable for soccer event prediction.

3.2.4.9 Qualitative Results

Figure 3.9 presents examples of recognition and pass prediction. The pass model needs to provide results for each pair of objects, but for presentation purposes, we choose the two most likely pairs. In Fig. 3.9a, the ball has left the passer and is moving toward the receiver but has not yet arrived. We use the data from this phase to train the pass prediction model. In Fig. 3.9b, the ball is already at the feet of the receiver, and the data from this phase are used to train the pass recognition model. These two phases have different features, leading to different behaviors in the prediction and recognition models. Figure 3.9c demonstrates that the shot recognition model can easily identify completed shots.

Figure 3.10 showcases several examples of intention prediction. In the first row of plots, the model initially predicts a high probability of a pass between the player with the ball and both teammates due to their proximity to the ball carrier and low defensive interference. As the defenders move up to impede the pass, the likelihood of the ball carrier passing in front of them decreases. Finally, when the ball carrier kicks the ball out, the model can identify that they pass the ball to a teammate on their right. The second row of plots shows an example of successive passes, where the model recognizes a change in the ball carrier after the initial ball is played to a teammate and predicts that the ball carrier will next pass to a teammate at the edge

Fig. 3.10 Examples of the intention prediction results on the SkillCorner dataset. The first and second rows show two passing events. The third row shows a shooting event. Players framed by white bounding boxes are ball carriers and players framed by red bounding boxes are possible receivers. The number above a white box represents the possibility whether the player is shooting and the number above a red box represents the possibility whether the ball carrier is passing to the player

of the court. The third row of plots presents an example of a final shot by the ball carrier, where the model reasonably predicts that the ball carrier is still likely to pass when they receive the ball, and the probability of predicting a shot increases as the ball carrier continuously adjusts their movement in front of the penalty area.

It is important to note that the model predictions do not add up to 1 because we use a sigmoid function instead of softmax to predict events. Softmax normalizes the predictions into probability form, but at a given moment, there are many reasonable choices, and the probabilities of these choices compete with each other. For example, if there are three well-positioned teammates, the ball carrier's probability of passing to each of them is not higher than 1/3. Using the sigmoid function to normalize the probability of an event assumes that all events are independent of each other and ignores their correlation, which is not a reasonable approach. However, graph neural networks themselves have the ability to utilize the global context and thus model the interactions. Furthermore, the results of the sigmoid function appear to be better interpreted as they are biased toward prediction plausibility rather than absolute probability. Therefore, we employ the sigmoid function to normalize the prediction results.

3.2.5 Conclusion

In this chapter, we have developed a novel multimodal architecture called OARNet for soccer event prediction tasks. OARNet effectively utilizes video data and spatiotemporal data to capture trajectory features and motion features of objects in a soccer game. The model leverages graph neural networks to model and reason about the interactions between different objects in the game. Moreover, we have introduced multihead GRU attention to handle potential multi-type information on relations, enhancing the model's performance. Our experimental results on two public datasets demonstrate that the proposed model achieves competitive performance compared to other state-of-the-art methods in passing and shooting event recognition and prediction. We have conducted extensive experiments to validate the effectiveness of each proposed component.

It is important to note that the multimodal feature fusion method can be further improved, and additional types of features can be incorporated to enhance the model's performance. The attributes of a player can be categorized into inherent static attributes and dynamic attributes that change over time during a match, which can be leveraged to facilitate graph reasoning more effectively. We anticipate that the principles and techniques employed in this research can also be applied to other sport domains to enhance event prediction capabilities.

References

1. Alahi, A., Goel, K., Ramanathan, V., Robicquet, A., Fei-Fei, L., Savarese, S.: Social lstm: human trajectory prediction in crowded spaces. In: Proceedings of the IEEE Conference on Computer Vision and Pattern Recognition, pp. 961–971 (2016)
2. Beal, R., Norman, T.J., Ramchurn, S.D.: Artificial intelligence for team sports: a survey. Knowl. Eng. Rev. **34**, e28 (2019)
3. Bialkowski, A., Lucey, P., Carr, P., Yue, Y., Sridharan, S., Matthews, I.: Large-scale analysis of soccer matches using spatiotemporal tracking data. In: 2014 IEEE International Conference on Data Mining, pp. 725–730. IEEE, New York (2014)
4. Cho, K., Van Merriënboer, B., Gulcehre, C., Bahdanau, D., Bougares, F., Schwenk, H., Bengio, Y.: Learning phrase representations using RNN encoder-decoder for statistical machine translation. arXiv preprint arXiv:1406.1078 (2014)
5. Clevert, D.A., Unterthiner, T., Hochreiter, S.: Fast and accurate deep network learning by exponential linear units (elus). arXiv preprint arXiv:1511.07289 (2015)
6. Davies, J., Studer, R., Warren, P.: Semantic Web Technologies: Trends and Research in Ontology-Based Systems. Wiley, New York (2006)
7. Dosovitskiy, A., Beyer, L., Kolesnikov, A., Weissenborn, D., Zhai, X., Unterthiner, T., Dehghani, M., Minderer, M., Heigold, G., Gelly, S., et al.: An image is worth 16×16 words: transformers for image recognition at scale. arXiv preprint arXiv:2010.11929 (2020)
8. Giancola, S., Amine, M., Dghaily, T., Ghanem, B.: Soccernet: a scalable dataset for action spotting in soccer videos. In: Proceedings of the IEEE Conference on Computer Vision and Pattern Recognition Workshops, pp. 1711–1721 (2018)
9. Gori, M., Monfardini, G., Scarselli, F.: A new model for learning in graph domains. In: Proceedings. 2005 IEEE International Joint Conference on Neural Networks, 2005, vol. 2, pp. 729–734. IEEE, New York (2005)

10. Gudmundsson, J., Horton, M.: Spatio-temporal analysis of team sports. ACM Comput. Surv. (CSUR) **50**(2), 1–34 (2017)
11. Gupta, A., Johnson, J., Fei-Fei, L., Savarese, S., Alahi, A.: Social gan: socially acceptable trajectories with generative adversarial networks. In: Proceedings of the IEEE Conference on Computer Vision and Pattern Recognition, pp. 2255–2264 (2018)
12. Guttenberg, N., Virgo, N., Witkowski, O., Aoki, H., Kanai, R.: Permutation-equivariant neural networks applied to dynamics prediction. arXiv preprint arXiv:1612.04530 (2016)
13. Hamilton, W., Ying, Z., Leskovec, J.: Inductive representation learning on large graphs. Adv. Neural Inf. Proces. Syst. **30**, 1–11 (2017)
14. Hochreiter, S., Schmidhuber, J.: Long short-term memory. Neural Comput. **9**(8), 1735–1780 (1997)
15. Huang, C.L., Shih, H.C., Chao, C.Y.: Semantic analysis of soccer video using dynamic bayesian network. IEEE Trans. Multimedia **8**(4), 749–760 (2006)
16. Huang, S., Chen, W., Zhang, L., Xu, S., Li, Z., Zhu, F., Ye, D., Chen, T., Zhu, J.: Tikick: towards playing multi-agent football full games from single-agent demonstrations. arXiv preprint arXiv:2110.04507 (2021)
17. Jiang, W., Higuera, J.C.G., Angles, B., Sun, W., Javan, M., Yi, K.M.: Optimizing through learned errors for accurate sports field registration. In: Proceedings of the IEEE/CVF Winter Conference on Applications of Computer Vision, pp. 201–210 (2020)
18. Khan, A., Lazzerini, B., Calabrese, G., Serafini, L., et al.: Soccer event detection. In: International Conference on Image Processing and Pattern Recognition (IPPR), vol. 3 (2018)
19. Khaustov, V., Mozgovoy, M.: Recognizing events in spatiotemporal soccer data. Appl. Sci. **10**(22), 8046 (2020)
20. Kim, K., Grundmann, M., Shamir, A., Matthews, I., Hodgins, J., Essa, I.: Motion fields to predict play evolution in dynamic sport scenes. In: 2010 IEEE Computer Society Conference on Computer Vision and Pattern Recognition, pp. 840–847. IEEE, New York (2010)
21. Kipf, T., Fetaya, E., Wang, K.C., Welling, M., Zemel, R.: Neural relational inference for interacting systems. In: International Conference on Machine Learning, pp. 2688–2697. PMLR, New York (2018)
22. Kipf, T.N., Welling, M.: Semi-supervised classification with graph convolutional networks. arXiv preprint arXiv:1609.02907 (2016)
23. Kurach, K., Raichuk, A., Stańczyk, P., Zając, M., Bachem, O., Espeholt, L., Riquelme, C., Vincent, D., Michalski, M., Bousquet, O., et al.: Google research football: a novel reinforcement learning environment. In: Proceedings of the AAAI Conference on Artificial Intelligence, vol. 34, pp. 4501–4510 (2020)
24. Lee, J., Nam, D.w., Moon, S., Lee, J., Yoo, W.: Soccer event recognition technique based on pattern matching. In: 2017 Federated Conference on Computer Science and Information Systems (FedCSIS), pp. 643–646. IEEE, New York (2017)
25. Li, J., Ma, H., Zhang, Z., Tomizuka, M.: Social-wagdat: Interaction-aware trajectory prediction via wasserstein graph double-attention network. arXiv preprint arXiv:2002.06241 (2020)
26. Li, J., Yang, F., Tomizuka, M., Choi, C.: Evolvegraph: multi-agent trajectory prediction with dynamic relational reasoning. Adv. Neural Inf. Proces. Syst. **33**, 19783–19794 (2020)
27. Li, Y., Tarlow, D., Brockschmidt, M., Zemel, R.: Gated graph sequence neural networks. arXiv preprint arXiv:1511.05493 (2015)
28. Liu, Y., Zhang, S., Mu, X., Ding, Z., Schober, R., Al-Dhahir, N., Hossain, E., Shen, X.: Evolution of NOMA toward next generation multiple access (NGMA) for 6G. IEEE J. Sel. Areas Commun. **40**(4), 1037–1071 (2022). https://doi.org/10.1109/JSAC.2022.3145234
29. Meystel, A.M., Albus, J.S.: Intelligent Systems: Architecture, Design, and Control. Wiley, New York (2000)
30. Morra, L., Manigrasso, F., Canto, G., Gianfrate, C., Guarino, E., Lamberti, F.: Slicing and dicing soccer: automatic detection of complex events from spatio-temporal data. In: Image Analysis and Recognition: 17th International Conference, ICIAR 2020, Póvoa de Varzim, Portugal, June 24–26, 2020, Proceedings, Part I 17, pp. 107–121. Springer, Berlin (2020)
31. Narasimhan, M., Lazebnik, S., Schwing, A.: Out of the box: reasoning with graph convolution nets for factual visual question answering. Adv. Neural Inf. Proces. Syst. **31**, 1–12 (2018)

32. Ogiela, L.: Semantic analysis in cognitive ubias & e-ubias systems. Comput. Math. Appl. **63**(2), 378–390 (2012)
33. Osokin, D.: Real-time 2d multi-person pose estimation on cpu: Lightweight openpose. arXiv preprint arXiv:1811.12004 (2018)
34. Richly, K., Bothe, M., Rohloff, T., Schwarz, C.: Recognizing compound events in spatio-temporal football data. In: IoTBD, pp. 27–35 (2016)
35. Richly, K., Moritz, F., Schwarz, C.: Utilizing artificial neural networks to detect compound events in spatio-temporal soccer data. In: Proceedings of the 2017 SIGKDD Workshop MiLeTS, Halifax, NS, Canada, pp. 13–17 (2017)
36. Sheu, P.C., Kitazawa, A., Ishii, C., Kaneko, K., Xie, F.: From semanticobjects to structured natural language. Int. J. Semant. Comput. **1**(03), 359–375 (2007)
37. Shi, X., Chen, Z., Wang, H., Yeung, D.Y., Wong, W.K., Woo, W.c.: Convolutional LSTM network: a machine learning approach for precipitation nowcasting. Adv. Neural Inf. Proces. Syst. **28**, 1–9 (2015)
38. Simonyan, K., Zisserman, A.: Very deep convolutional networks for large-scale image recognition. arXiv preprint arXiv:1409.1556 (2014)
39. Tavassolipour, M., Karimian, M., Kasaei, S.: Event detection and summarization in soccer videos using bayesian network and copula. IEEE Trans. Circuits Syst. Video Technol. **24**(2), 291–304 (2013)
40. Vaswani, A., Shazeer, N., Parmar, N., Uszkoreit, J., Jones, L., Gomez, A.N., Kaiser, Ł., Polosukhin, I.: Attention is all you need. Adv. Neural Inf. Proces. Syst. **30**, 1–11 (2017)
41. Veličković, P., Cucurull, G., Casanova, A., Romero, A., Lio, P., Bengio, Y.: Graph attention networks. arXiv preprint arXiv:1710.10903 (2017)
42. Wang, Y.: On cognitive informatics. Brain and Mind **4**, 151–167 (2003)
43. Wang, Y.: Using process algebra to describe human and software behaviors. Brain and Mind **4**, 199–213 (2003)
44. Wang, Y.: On concept algebra for computing with words (CWW). Int. J. Semant. Comput. **4**(03), 331–356 (2010)
45. Wang, Y.: On formal and cognitive semantics for semantic computing. Int. J. Semant. Comput. **4**(02), 203–237 (2010)
46. Wickramaratna, K., Chen, M., Chen, S.C., Shyu, M.L.: Neural network based framework for goal event detection in soccer videos. In: Seventh IEEE International Symposium on Multimedia (ISM'05), pp. 8–pp. IEEE, New York (2005)
47. Xu, D., Zhu, Y., Choy, C.B., Fei-Fei, L.: Scene graph generation by iterative message passing. In: Proceedings of the IEEE Conference on Computer Vision and Pattern Recognition, pp. 5410–5419 (2017)
48. Xu, K., Ba, J., Kiros, R., Cho, K., Courville, A., Salakhudinov, R., Zemel, R., Bengio, Y.: Show, attend and tell: neural image caption generation with visual attention. In: International Conference on Machine Learning, pp. 2048–2057. PMLR, New York (2015)
49. Xu, K., Hu, W., Leskovec, J., Jegelka, S.: How powerful are graph neural networks? arXiv preprint arXiv:1810.00826 (2018)
50. Yang, J., Lu, J., Lee, S., Batra, D., Parikh, D.: Graph r-cnn for scene graph generation. In: Proceedings of the European Conference on Computer Vision (ECCV), pp. 670–685 (2018)
51. Yue, Z., Broich, H., Seifriz, F., Mester, J.: Mathematical analysis of a soccer game. Part I: individual and collective behaviors. Stud. Appl. Math. **121**(3), 223–243 (2008)
52. Yue, Z., Broich, H., Seifriz, F., Mester, J.: Mathematical analysis of a soccer game. Part II: Energy, spectral, and correlation analyses. Stud. Appl. Math. **121**(3), 245–261 (2008)
53. Zadeh, L.A.: Some reflections on soft computing, granular computing and their roles in the conception, design and utilization of information/intelligent systems. Soft. Comput. **2**, 23–25 (1998)
54. Zhu, W., Han, Y., Lu, J., Zhou, J.: Relational reasoning over spatial-temporal graphs for video summarization. IEEE Trans. Image Process. **31**, 3017–3031 (2022)

Chapter 4
Structural Coding

4.1 Sketch Graph Representation for Multimedia Computational Communications: A Learning-Based Method

Multimedia computational communications in the context of 6G have the potential to significantly improve transmission efficiency by incorporating intelligent computation into the communication process. This intelligent communication architecture encompasses multimedia representation, coding, transmission, and other aspects from a semantic perspective. The core component of this architecture is multimedia semantic representation, which primarily aims to reduce the amount of multimedia data. In this chapter, we propose the use of sketch graphs as an effective representation of images, capturing pixel variations, geometric feature distribution, and structural information. Sketch graphs hold promise for applications in multimedia computational communications. Specifically, we develop a learning-based method that utilizes deep neural network (DNNs) to extract sketch graphs through edge detection, sketch point detection, and sketch line detection. Moreover, we design an end-to-end extraction method that achieves real-time processing. Experimental results on several datasets demonstrate the superior performance of our approach in terms of classification and generation tasks. Image classification results on the HumanSketch, ImageNet, and Caltech datasets indicate that our method produces sketch graphs with better descriptive capabilities compared to traditional methods and other conventional image compression techniques. Furthermore, image generation results on the Cityscapes dataset highlight the potential of the sketch-graph-based image compression codec.

X. Tao et al., *Wireless Multimedia Computational Communications*,
Wireless Networks, https://doi.org/10.1007/978-3-031-64155-8_4

4.1.1 Introduction

As multimedia capture devices continue to advance and the demand for multimedia services grows, efficient multimedia communications involving large data volumes have received significant attention from researchers. The concept of computational communications has emerged as a means to improve the transmission efficiency of multimedia content by leveraging AI techniques. On the sender side, high-level semantics are extracted with AI methods to reduce the amount of transmitted data. On the receiver side, the semantics are reconstructed into multimedia data via deep reconstruction models or employed for downstream tasks. Throughout this process, the focus shifts from pixel-level elements to semantic representations, maximizing the retention of meaningful information. The ability to perform efficient semantic representations of multimedia content, such as images and videos, becomes a key factor influencing the performance of multimedia computational communications.

In the field of image semantic representations, object categories are one of the most widely utilized and effective types of semantic information. By capturing category features of the main target through semantic segmentation or scene graph generation methods, the amount of transmitted data can be greatly reduced. However, reconstructing images solely based on category labels (e.g., image reconstruction based on scene graphs [40]) introduces strong uncertainty, and the synthesized results may contradict the actual properties of the target. Reconstruction based on semantic segmentation maps [39] ensures the correctness of the target outlines but still lacks the characterization of the internal textures, which are also important for identity recognizing. Therefore, combination with other lower-level auxiliary information becomes a more viable approach for accurately characterizing target attributes without substantially increasing data volume.

The sketch graph model, proposed by Marr [50], is a sparse representation of images. Cheng-en Guo [31] developed the primal sketch and compressed the main edges and textures by linear segment fitting and dictionary coding, respectively, which achieved significant compression effects on grayscale images. Shuai Yang [79] combined sketch graph and color sampling, achieving superior performance compared with traditional image compression algorithms. However, current sketch graph extraction methods basically rely on the greedy search strategy. Although they can effectively fit the main edges of an image, their threshold-based search approach lacks flexibility in meeting diverse bandwidth requirements, making it challenging to adapt to different scenarios.

Therefore, we propose a new sketch graph representation model, deep sketch graph (DSG), and design a DNN-based method for sketch extraction. We model sketches as collections of points and lines using DNNs, enhancing the representation performance of sketches. The main contributions of this chapter are listed as follows:

- We introduce DSG as a model to describe pixel variations, geometric feature distribution, and structural information of images by means of a point-line collection, which enhances the representation capability of sketch graphs.

- We propose a DNN-based sketch graph detection model that enables parallel extraction of sketch points and lines, achieving efficient extraction with graphics processing unit (GPU) acceleration. Our experiments demonstrate real-time detection on large-scale images.
- We propose a compressed coding method for DSGs to further reduce data volume while better-preserving sketch graph features. This method achieves superior image characterization performance compared to traditional image representation methods with comparable data sizes.

4.1.2 Related Works

4.1.2.1 End-to-End Multimedia Computational Communications

Traditional communication systems are usually composed of several cascaded modules, such as source encoding, channel encoding, modulation, and corresponding inverse processes, which are individually designed and optimized. With the advancements in deep learning, end-to-end multimedia computational communications, based on joint source-channel design, have achieved improved performance in tasks like text and image transmission. In addition to automatically learned features, researchers have also imposed artificial constraints on network outputs to tailor the model to specific tasks.

Among these artificial constraints, the sketch graph model shows promise due to its sparse properties, which facilitate the reduction of transmitted data, enabling communication in limited bandwidth scenarios.

4.1.2.2 Image Semantic Representations

Image semantic representation refers to the extraction of the most relevant semantic information from images in specific contexts and is usually accompanied by advanced feature extraction techniques, such as identity and category recognition for semantic segmentation tasks, detection of main targets and relationships for scene graph generation tasks, and description of identities in a scene and their relationships in natural language for image caption tasks.

While traditional sketch graphs employ a pixel-based representation method that abstracts and compresses the main edges of an image, the deep sketch graph model proposed in this chapter combines pixel representation with additional semantic information, thereby reducing overall data volume while providing a certain level of detail representation.

4.1.2.3 Edge Detection and Sketch Graph

The current mainstream edge detection methods mainly include feature design–based methods [9, 43, 57] and deep learning–based methods [33, 66, 74]. Traditional sketch extraction algorithms typically involve geometric abstraction of image edges using iterative or traversal techniques, primarily to eliminate spatial redundancy in edge images. Cheng-en Guo [31] designed an iterative algorithm to extract the sketch graph by fitting consecutive image edges with straight-line segments, but encountered strong discontinuities. Jie Wu [72] designed a recursive algorithm that traversed all non-crossing edges and extracted the sketch graph by fitting segments to curved edges based on the Ramer-Douglas-Peucker algorithm, which is greatly affected by the edge detection results. In addition, Shuai Yang [79] extracted sketch graphs by using AutoTrace software for edge lines in SVG encoding format, but this approach relied on non-open-source software, limiting its editability and scalability.

Given the advancements in deep learning technology, there is a natural and pressing need to explore deep learning–based methods for extracting sketch graphs and integrating them with other semantic information.

4.1.3 Learning-Based Sketch Graph Model

4.1.3.1 Overall Framework

Our system consists of three parts: edge detection, sketch point detection, and sketch line detection. In contrast to the primal sketch graph, which defers the representation of textual information to neural networks, the deep sketch graphs proposed in this chapter comprise two main components: sketch points and sketch lines. Building on the concept of the primal sketch, the sketch points are classified into three categories: cross points, corner points, and end points, representing points connected to three or more sketch points, two sketch points, and one sketch point, respectively. Sketch lines reflect the connections between two sketch points. We primarily employ convolutional neural networks (CNNs) for the detection of sketch graphs. The general structure is as follows: first, CNNs are used for edge detection, followed by non-maximal suppression (NMS) to obtain single-pixel edges; second, sketch points are extracted; and lastly, sketch lines are detected by combining sketch points with the single-pixel edges using feature sampling. The sketch graph comprises sketch points and sketch lines (i.e., connection matrix of each sketch point pair), based on which we can produce a sketch image and conduct different downstream tasks.

4.1.3.2 Data Preparation

We expect the sketch points and lines to contain not only the boundaries of different targets but also the main textures within the targets. To achieve this, we utilize the widely used BSDS500 dataset [3], an edge detection dataset containing 500 diverse natural images with 5–6 manually labeled edges per image, for training our networks. First, we train the CNN-based edge detection network using the images and edge annotations in the training and validation sets. Second, we apply graphical correction to the edges and employ the sketch graphs detected by the sketch extraction method in [72] as the ground truth for the training process.

4.1.3.3 Network Design and Inference

For edge detection, we use the hourglass structure [55] for feature extraction, and the output is then processed by a convolution layer with a single convolution kernel, followed by the sigmoid activation function, to obtain the probability of each pixel belonging to edges. The edge probability map is further processed through an NMS network to obtain the single-pixel edge map. The design of the NMS network refers to the method employed in PidiNet [66] and is realized in the Pytorch framework to enable GPU acceleration.

For sketch point detection, deep features are extracted from the edge map using the hourglass module. A convolution layer with five convolution kernels, followed by a softmax layer, is utilized to determine the category of each pixel. The output consists of five channels representing the probability of each pixel belonging to intersection points, inflection points, end points, non-sketch points, and non-edge points.

For sketch line detection and connection detection, features are extracted from the single-pixel edge using an hourglass module. Subsequently, S features are uniformly sampled from the concatenation of each point pair. For example, features at locations $\{(x_1 + s \cdot \Delta x, y_1 + s \cdot \Delta y_s), s = 0, 1, \cdots, S-1\}$ are sampled as features for the point pair $\{(x_1, y_1), (x_2, y_2)\}$, where $\Delta x_s = \frac{x_2 - x_1}{S-1}$, and $\Delta y_s = \frac{y_2 - y_1}{S-1}$. The sampled features are then processed through the CNNs, and the probability that each point pair is connected is obtained using the sigmoid activation function. The overall framework is shown in Fig. 4.1.

The inference process differs from the aforementioned description. For the detection of sketch points, based on the scale-sensitive property, we adopt a multiscale detection mechanism, which scales the input edges at different scales and fuses the extracted sketch points to obtain a more accurate detection.

Regarding connectivity detection, the computation required for numerous sketch points is immense since the number of feature samples is proportional to the square of the sketch point number. To alleviate this, we employ a chunking-summarization mechanism. Initially, since connected sketch points within a local region tend to be grouped, we perform a simple spatial division of all sketch points and detect sketch connectivity for each group. For the edge lines missed as a result of two separated sketch points in different blocks, we gather all unsaturated sketch points again for sketch line detection and obtain the final sketch graph.

Furthermore, to enhance the robustness and accuracy of the grouping-based detection, we perform connection revision after the detection. For example, when

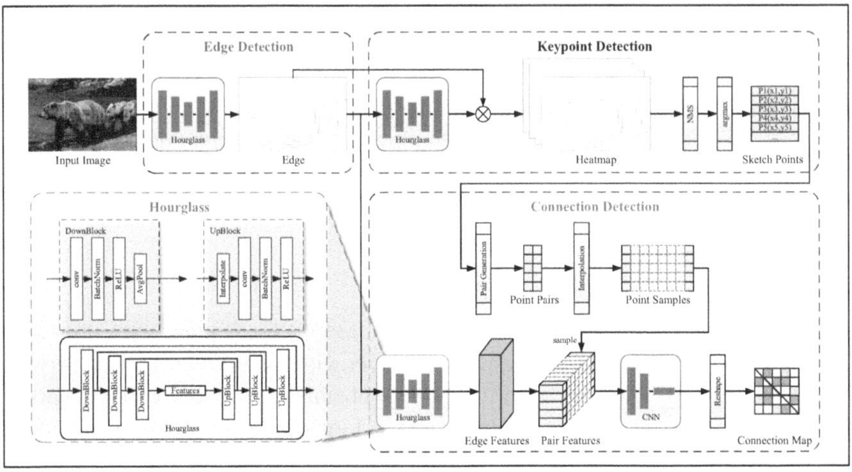

Fig. 4.1 Framework of sketch graph detection. We primarily utilize hourglasses with separate parameters as the backbones for feature detection in each module and use sampled features for connection detection. The final sketch graph consists of 2D arrays representing sketch points and the adjacency matrix of all vertices

three sketch points are interconnected and almost arranged in a line, the connection of the pair with the largest distance is canceled. Additionally, if a sketch point has a current degree exceeding the theoretical value, the weakest connection may be canceled based on the location of the connected points. Subsequently, weak connections between two unsaturated points are strengthened to compensate for the likelihood of missing connections.

4.1.3.4 Sketch Graph Modification and Compression Encoding

To enable the sketch graph to better fit the edges with a smaller data volume, we design a modification and compression algorithm. The overall procedure includes redundant connection elimination, missing connection supplementation, and sketch graph compression.

Depending on the reasons for the redundancy, we divide the redundant sketch lines into three categories: incorrectly drawn lines, repeatedly drawn lines, and nonmaximal lines. These categories represent connections leading to incorrect sketch lines, connections already covered by others, and connections resulting in thick edges. We assign zeros to the corresponding elements in the connection matrix to eliminate redundant sketch lines. Subsequently, we draw the sketch according to the sketch graph and supplement the disconnected sketch pairs to achieve a better fit with the actual edges.

For sketch graph compression, we search for the sketch points among all corner points that have the least impact on the overall sketch and iteratively remove such points to obtain more streamlined sketch graphs with smaller data volumes.

We design a lossless encoding algorithm based on the sparse nature of sketch graphs and further reduce the data required to represent a sketch graph. For sketch points, we represent each two-dimensional location with bits of length $\lceil \log_2 W_{\text{image}} \rceil + \lceil \log_2 H_{\text{image}} \rceil$, where W_{image} and H_{image} represent the width and height, respectively, of a sketch image, and both are also encoded into the streams. For the connection matrix composed of 0's and 1's, we rearrange all the sketch points such that points on a segmentation line are arranged in order. This rearrangement operation ensures most of the 1's on the subdiagonal of the matrix. Then, we perform XOR operations on the connection matrix and a matrix with all 1's on the subdiagonal and 0's for the other elements and encode the location of the resulting 1's in the output matrix to represent the connection matrix. As sketch points are usually connected to form lines, our encoding method can greatly reduce the data volume.

In contrast, we apply compression encoding on sketch graphs extracted using traditional methods, where sketch points and sketch lines are represented as arrays with the same bits of length and series of 0's and 1's, respectively. The key difference lies in the encoding of the location of 1's in the XOR result of the connection matrix. Traditional methods encode the entire matrix with 1 bit for each element.

4.1.4 *Experiments, Results, and Analysis*

4.1.4.1 Loss Functions and Training Settings

The proposed network is mainly oriented to classification tasks, so binary cross-entropy (for edge detection and sketch line detection) and multivariate cross-entropy (for sketch point detection) are utilized as loss functions. Considering the unbalanced sample characteristics of edges and sketch points, a weighted cross-entropy is designed to train the network based on the sample distribution characteristics of the dataset. A stepped learning rate adjustment strategy is adopted for the training process, and the Adam optimizer is used for updating the network parameters.

4.1.4.2 Comparison Algorithms and Metrics

In this chapter, the performance of the proposed deep sketch graph is analyzed from two perspectives: the effectiveness of the edge-based sketching method and the performance of the sketch graph as a new image representation form, which will be compared with other sketch extraction methods as well as other image representation or compression methods.

First, we compare the sketching performance of different algorithms using the same set of edge images. For the quality evaluation of sketch graphs, there is a lack of methods for comparison, and the identified algorithms include those of primal sketch [31] and SAR sketch [72]. Considering the accessibility of codes, we only conduct a comparison with the latter algorithm, and a fixed threshold for fitting error is set for SAR sketch to extract sketch graphs across the entire dataset.

Second, we compare the compression performance of different image representation algorithms using a natural image. To evaluate image characterization performance, we mainly compare DSG sketches with semantic segmentation maps, SAR sketches, traditional image compression methods such as JPEG, WEBP, JPEG2000 (J2K), BPG, and the learning-based image compression method HiFiC [52].

Additionally, we also employ graph compression methods k^2-tree [8] and re-pair [10] to compress the proposed sketch.

4.1.4.3 Evaluation Dataset and Metrics

The evaluation methods for the quality of sketch graphs are limited, with existing methods often focused on image reconstruction or the implementation of downstream tasks (classification and retrieval, etc.) based on sketches. Therefore, in this chapter, the quality of sketch graphs is represented by their performance in different tasks. With reference to current mainstream sketch-based computer vision tasks, we perform sketch classification on the QuickDraw [32] and HumanSketch datasets [24] for object edges, as well as ImageNet [16] and Caltech101 [26] for natural images. The performance of different sketches with the same classifier is measured to reflect the quality of the sketches themselves, and comparisons with traditional image compression methods are conducted to evaluate how well the sketch graphs retain object attributes. The ResNet50 classifier is utilized in this chapter, and mean accuracy (mAcc, the average classification accuracy across multiple instances in a dataset) and mean bits per pixel (bpp, the average number of bits used to encode one pixel) are two indicators for quantitative evaluation. The classification task is performed in 50 randomly selected categories.

In addition, the pixel-to-pixel network (pix2pix) [39] is utilized to reconstruct the image based on different image representation forms on Cityscapes [14] and the mean intersection over union (mIoU, the average ratio of the intersection between the predicted segmentation and the ground truth segmentation over the union of the two) is examined to reflect the ability of different compression methods to retain the target categories. We also compare with JPEG, WEBP, J2K, BPG, and HiFiC. The pix2pix models are trained with the same hyperparameters and structures to ensure fairness.

4.1.4.4 Performance and Analysis

Figure 4.2 shows the performances of different sketch extraction methods for the classification task. Subfigures (a) and (b) demonstrate that DSG sketches achieved higher accuracy than SAR sketches with the same bpp, indicating that the proposed method has stronger characterization ability and preserves more characteristics

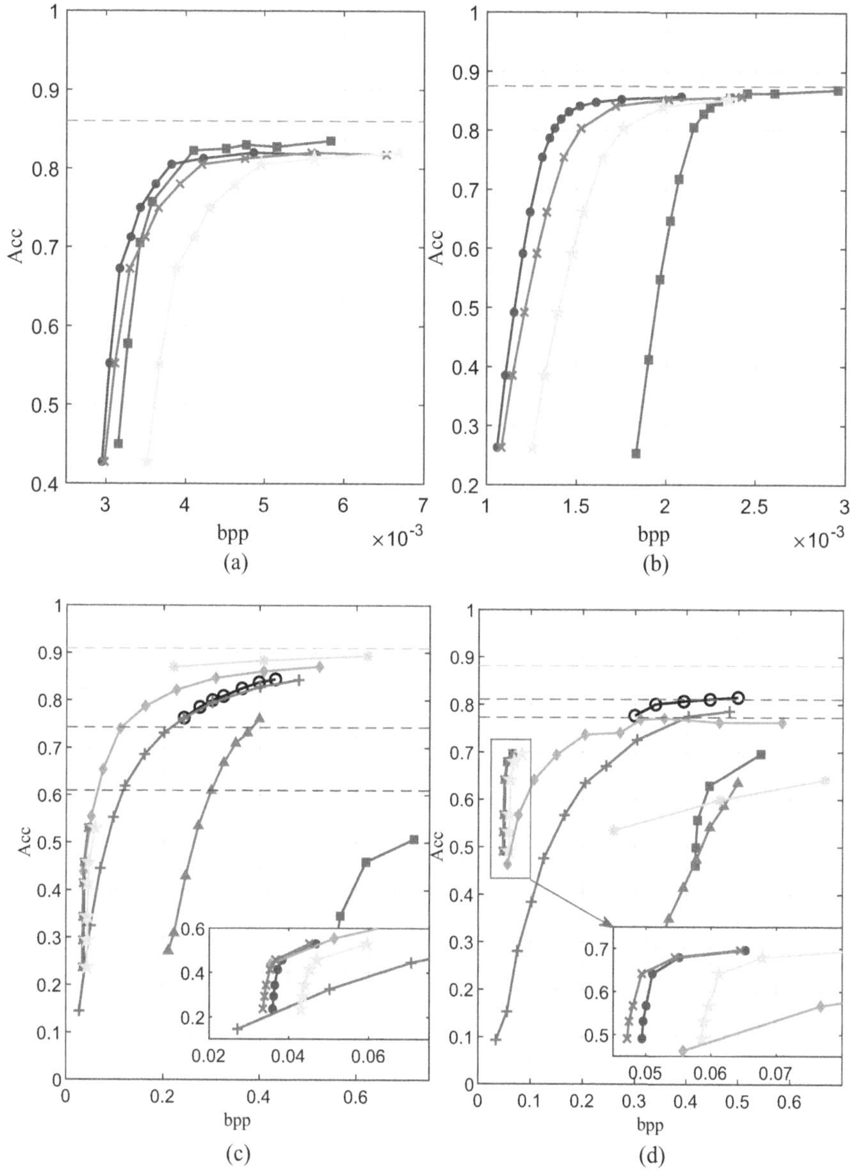

Fig. 4.2 Performance on classification task. Magenta dashed lines, orange dashed lines, red dashed lines, and blue dashed lines represent the classification performance on original sketches, original images from the dataset, edges extracted by the Canny operator, and PidiNet, respectively. When comparing artificial datasets (**a**) and (**b**) with datasets of natural images (**c**) and (**d**), it is evident that the performance of sketches is strongly influenced by the complexity of the image backgrounds. Furthermore, the comparison between (**c**) and (**d**) indicates that simpler backgrounds are more conducive to sketch representation. (**a**) HumanSketch. (**b**) QuickDraw. (**c**) ImageNet. (**d**) Caltech

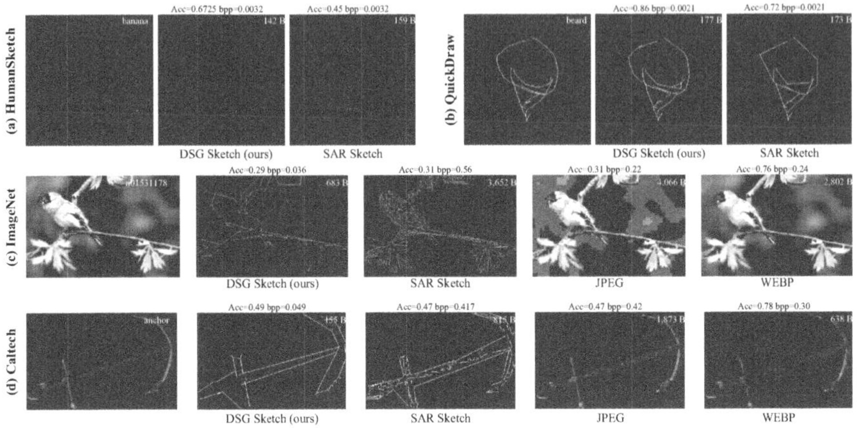

Fig. 4.3 Sketch graph examples by SAR and DSG sketches and compression results of JPEG and WEBP. A fixed threshold (for sketch graph extraction) or compression factor (for image compression) is applied to the entire dataset, and the average bpp and classification accuracy are calculated for a plotted point in Fig. 4.2. Here, a selection of samples from the dataset is displayed, with the bpps of each image marked. Subfigures (**a**) and (**b**) present the results under the same amount of data, while (**c**) and (**d**) show the results under the same performance level since the data amount of the image compression algorithm cannot reach a comparable level with others. Due to space constraints, results with compression methods JPEG2000, BPG, and HiFiC are not displayed

related to the target category. In addition, the achieved bpp reaches an order of magnitude of 10^{-3}, highlighting the potential application of compression for images with simple outlines and textures. Subfigures (c) and (d) reveal that for natural images with complex textures and backgrounds, our method surpasses JPEG, J2K, BPG, and SAR sketch in terms of classification accuracy and data amount, although the compression performance is slightly inferior to k^2-tree. While our method's highest accuracy falls short of WEBP and HiFiC, it still achieves a relatively low bpp with an order of magnitude of 10^{-2}. Figure 4.3 provides examples of sketches obtained using different sketch graph extraction and image compression methods.

Performances of different image representation forms (i.e., segmentation maps, edges, and sketch graphs) and compression methods are also analyzed. Some examples based on different image representation methods are shown in Fig. 4.4a, while quantitative results are shown in Fig. 4.4b. It can be seen from subfigure (b) that the proposed DSG exhibits clear advantages over traditional sketch graph as well as graph compression methods k^2-tree and re-pair. For example, DSG representations reach an mIoU of 0.1235 at an bpp of 0.041, whereas JPEG and WEBP attain a comparable performance at bpps of 1.04 and 0.43, respectively. Moreover, as shown in subfigure (b), our DSG sketches outperform SAR sketches in terms of performance at a lower bpp. Although DSG is inferior to semantic segmentation maps, we could still see the potential in image representation, with gaps filled by other methods such as residual transmission.

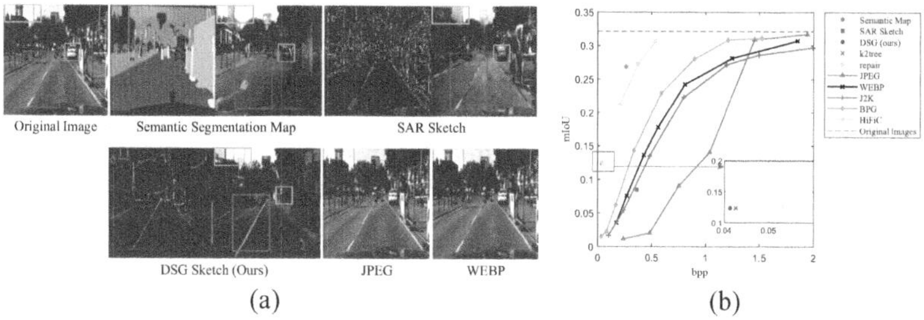

Fig. 4.4 Performance on reconstruction task. (**a**) shows the reconstruction results from semantic segmentation maps, sketch graphs, edges, and JPEG and WEBP methods at approximately 0.12 for mIoU. Compared to semantic segmentation maps, sketches better preserve the textual characteristics of the reconstructed images (as indicated by the red boxes marking lane lines). However, the absence of certain semantic information leads to biased reconstruction results in terms of target categories (as indicated by the blue boxes marking buildings) and the appearance of small targets (as indicated by the green boxes marking a car). (**b**) presents the quantitative results, where semantic segmentation maps were losslessly compressed using the online API tinypng

In terms of computational complexity, the single-shot inference of the proposed network is real-time under GPU acceleration, surpassing traditional algorithms [72]. However, with the introduction of multi-scale and later correction, the detection time significantly increases, exceeding that of traditional algorithms.

4.1.5 Conclusion, Discussion, and Future Work

In this work, we propose a learning-based method for extracting sketch graphs from natural images. We are the first to employ neural networks to extract highly structured and sparse sketch graphs, validating their performance from multiple perspectives. The innovation of our work lies in the following aspects: (i) experiments on sketch image datasets (HumanSketch and QuickDraw) demonstrated that our learning-based method achieved better classification performance compared to traditional sketch graph extraction and compression methods; (ii) experiments on natural image datasets (ImageNet, Caltech, and Cityscapes) revealed that the deep sketch graph can be considered a new form of image representation that preserves both semantic and textual information. Although there is some loss of semantic information during the sketching process, combining sketch graphs with certain semantic labels could potentially address this issue, which is a focus of our future research.

4.2 Panoramic Image Generation: From 2D Sketch to Spherical Image

The emergence of 360-degree videos/images, also known as omnidirectional or panoramic videos/images, has gained significant importance in areas such as virtual reality (VR). As a result, there is a growing need for corresponding image generation algorithms. However, existing image generation models mainly focus on 2D images and do not take into account the spherical structure of panoramic images. In this chapter, we propose a panoramic image generation method called spherical generative adversarial networks (SGANs), which is based on spherical convolution and generative adversarial networks (GANs). We adopt a sketch map as the input, which is a concise geometric structure representation of the panoramic image and occupies only approximately 7% of the pixels for a 583×1163 image. Through adversarial learning, realistic-looking, plausible, and high-fidelity spherical images can be obtained from the sparse sketch map. In particular, we build a dataset of sketch maps using a visual computation–based sketching model. Then, by optimizing SGANs with GAN loss, feature matching loss, and perceptual loss, realistic textures and details are recovered progressively. On the one hand, the utilization of the sparse sketch map as input rather than the denser inputs, such as the textures and colors, constitutes an improvement. On the other hand, spherical convolution helps to address space-varying distortions caused by the planar projection. We conduct extensive experiments on some public panoramic image datasets and compare the results with state-of-the-art techniques to validate the superior performance of the proposed approach.

4.2.1 Introduction

With the rapid development of multimedia technologies and applications such as smart tourism, virtual reality, and unmanned driving, panoramic images have gained attention due to their full-view features. The panoramic image is also called a 360-degree or omnidirectional image [76]. Considerable research efforts in 360-degree image processing have been witnessed in recent years. In general, 360-degree image processing is a new multimedia technology for improving the QoE. Panoramic image generation is essential in studying the QoE of viewing panoramic images. However, the generation model of a panoramic image is significantly different from that of a traditional image. The main difference is that a panoramic image offers an immersive and interactive viewing experience, as the viewers are able to freely move their heads in the range of $360° \times 180°$ to access different viewpoints. Therefore, the generation models and corresponding applications for panoramic images deserve to receive more attention [48].

The problem of image generation is essentially an inverse problem, which can be formulated as follows:

$$\mathbf{M} = f(\mathbf{y}_{ms}) + \boldsymbol{\xi} \tag{4.1}$$

where $\mathbf{M}$ represents the generated image, $\mathbf{y}_{ms}$ is a measurement vector, $f(\cdot)$ is the function that recovers the original signal from the measurements, and $\boldsymbol{\xi}$ denotes the generation error. According to information theory, the more measurements we observe, the better the quality of images we can thereby generate [75]. The Nyquist sampling theorem states that if we know the highest frequency of the signal $fmax$, the sampling theorem provides the lowest sampling frequency $fs > 2fmax$ that guarantees a fully reconstructed signal [73]. Transform coding [70] is a traditional image reconstruction method that achieves signal compression by storing only the large transform coefficients of the signal. During signal reconstruction, it is only necessary to set the coefficient that is not stored to 0, and then reconstruct the signal through the corresponding inverse transform. Afterward, compressed sensing [21] and sparse representation [27] are employed to reconstruct signals at sampling rates much lower than the Nyquist frequency. These methods are based on dictionary learning and sparse representation theory, utilizing sparse coefficients for signal reconstruction.

Recently, GANs [29] have been shown to be capable of generating visually satisfactory images, making it difficult even for humans to discern between generated and authentic images. In addition, GANs generate images with a relatively small amount of data, such as a noise vector or semantic label map [84]. Therefore, GANs have achieved promising results in natural image generation across various datasets, such as the VGG face dataset [15], and the ImageNet dataset. However, panoramic images differ fundamentally from natural images due to their distinct imaging mechanism. Specifically, panoramic images consist of spherical images captured from different viewpoints. When applying naive convolution, the spherical images need to be projected onto a planar space, which introduces space-varying distortions, as shown in Fig. 4.5. It should be noted that some distortions exist in the

(a)

Fig. 4.5 Projection of a spherical panoramic image onto a 2D plane. The left side shows the spherical image, while the right side shows the plane image. Distortions, such as those in the road and tree, are observed in the plane image

planar image, such as those on the road and trees. On the other hand, most existing works do not consider the structural aspects of images during the generation process (when generating from random noise or semantic maps), despite their significance to users' QoE.

Therefore, there is a need to develop specific generation methods for panoramic images. In this chapter, we propose a panoramic image generation method that automatically translates a 2D sketch into its spherical form. This approach represents a way to balance the conflict between the input data amount and the desired high quality. In particular, we build a dataset of sketch maps using a visual computation–based sketching model. Then, by optimizing SGANs with GAN loss, feature-matching loss, and perceptual loss, we progressively recover textures and details, ultimately generating a spherical panoramic image. The entire network is trained end-to-end, starting from the sparse contour to the spherical image. The framework of our proposed approach is illustrated in Fig. 4.6. With the contour of a panoramic image, our model can generate textures and details using statistical priors derived from a large dataset. From a communication system perspective, this approach significantly reduces the volume of transmitted data (from panoramic image to sketch map), as shown in Fig. 4.7. For a panoramic image of size 583×1163, the sketch map accounts for only approximately 7% of the total pixels, consisting of 44,861 sketch points. The main contributions of this chapter are twofold:

- We propose a generative adversarial network with spherical convolution for panoramic image generation, which reduces the distortions caused by 2D planar mapping.

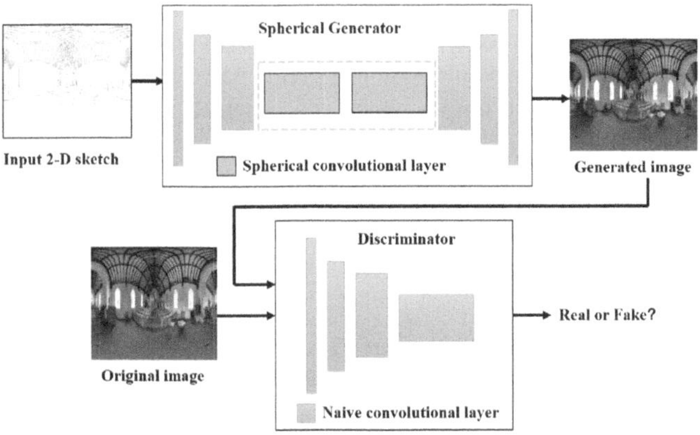

Fig. 4.6 Framework of the proposed approach with the paradigm of SGANs. Using a 2D sketch map as input, the spherical generator synthesizes the panoramic image. The discriminator distinguishes between the generated images and real images

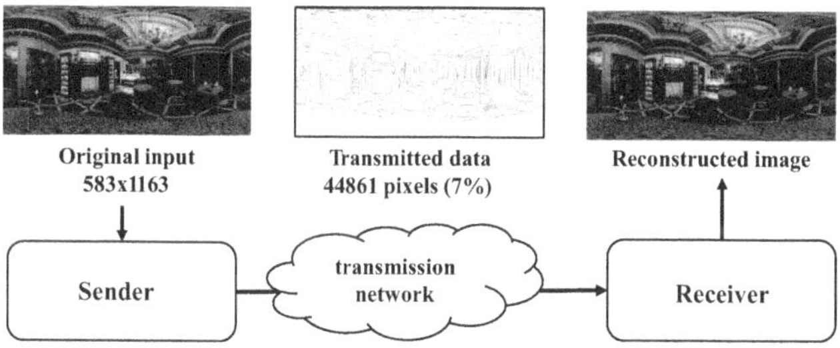

Fig. 4.7 A communication system utilizing the sketch representation. For the sender, it extracts the sketch map of the original signal/image, where sketch points only account for seven percent of the image pixels. For the receiver, it generates the reconstructed image using the transmitted sketch map, significantly reducing the amount of transmitted data

- We develop a contour-based generation method. By utilizing the sparse sketch map as input, we fill in textures and details and generate the spherical image by adversarial learning. This constitutes a significant improvement from the 2D sparse sketch map to the spherical image.

The rest of this chapter is organized as follows: Sect. 4.2.2 reviews related work on image generation. Section 4.2.3 describes the construction of the sparse sketch map for the panoramic image and presents the panoramic image generation model based on SGANs. Section 4.2.4 represents experimental results, and Sect. 4.2.5 concludes this chapter.

4.2.2 *Related Work*

With the growing demand for high-quality images in visual applications, image generation has become a prominent topic in computer vision and image processing. Previous methods for panoramic image generation mainly focus on the stereo imaging system, including sensors and input-output devices [41]. In contrast, our work approaches panoramic image generation from a signal-processing perspective. In this section, we briefly review relevant literature on image reconstruction/generation, covering three main aspects: transform coding, compressed sensing, and deep learning.

4.2.2.1 Transform Coding–Based Reconstruction/Generation

Linear transformation of an image involves linearly transforming the image from the spatial domain to another transform domain. Following the transformation, image compression can be achieved by applying appropriate quantization and entropy coding methods based on rate-distortion theory. As long as the distortion of the reconstructed image is within the allowable range, any compression ratio can be used with greater flexibility. Therefore, in previous research on image processing, image reconstruction based on transform coding is the most common algorithm. For instance, in [54], the authors explored a set of orthogonal moment functions based on discrete Tchebichef polynomials and demonstrated the effectiveness of Tchebichef moments as feature descriptors in image reconstruction. Elshoura et al. [25] proved that Tchebichef moments outperformed Legendre and Zernike moments in image reconstruction, regardless of the nature of the image used, by analyzing image reconstruction accuracy in the frequency domain. In [37], Hong et al. proposed an adaptive reconstruction method for loss blocks in images, where the discrete cosine transform (DCT) coefficients were recovered by adaptively selecting available neighboring blocks and performing interpolation. The experimental results demonstrated successful image quality restoration with low computational complexity. Gunturk et al. [30] proposed a random framework that efficiently used quantitative operations, transform-domain statistics, and additive sensor noises for image super-resolution reconstruction. Wavelet transform, a time-frequency analysis method, has shown excellent performance in both the time and frequency domains. Given a continuous signal $f(t) \in L^2(R)$, the wavelet transform can be expressed as:

$$f(t) = \sum_{k} \beta_{j_0,k}(t)\varphi_{j_0,k}(t) + \sum_{j=j_0}^{\infty} \sum_{k} \alpha_{j,k}\Psi_{j,k}(t) \tag{4.2}$$

where $\varphi_{j_0,k}(t)$ represents the scaling function and $\Psi_{j,k}(t)$ denotes the wavelet function. The emergence of wavelet transform has made image processing more convenient [2] and has become a classic method for image reconstruction. In [65], Starck et al. Starck et al. performed multi-resolution deconvolutions on interference images using wavelet coefficients and reconstructed images of two evolutionary stars from infrared speckle interferometry. The discrete wavelet transform has also been applied to SAR image generation problems [6].

4.2.2.2 Compressed Sensing–Based Reconstruction/Generation

In contrast to transform coding methods, several compressed sensing–based approaches have been proposed for image reconstruction algorithms. Compressed sensing [21] can sense signals at a rate much lower than that of the Nyquist sampling theorem, while retaining the basic information contained in the signal allows signal sensing at rates much lower than the Nyquist sampling theorem while retaining essential information contained in the signal. Image reconstruction based on compressed sensing aims to estimate the original image from compressed observations, often involving complex calculations to solve highly nonlinear optimization reconstruction problems. The goal of compressed sensing is to recover the signal z from its measurements y. The basic model for compressed sensing image reconstruction is formulated as follows:

$$z^* = \arg\min_{z} \|z\|_0 \quad s.t. \quad \|y_{ms} - \Phi z\|^2 \leq \varepsilon \tag{4.3}$$

where $\|\cdot\|_0$ represents the l_0 norm and is used to calculate the number of nonzero elements in the signal. Φ^{sen} is a sensing matrix that quantifies the signal's sparsity. When the signal's sparsity is known, the reconstruction can be performed by solving the following model:

$$z^* = \arg\min_{z} \|y_{ms} - \Phi z\|^2, \quad s.t. \quad \|z\|_0 \leq S_K \tag{4.4}$$

where S_K represents the sparsity of the signal.

The optimization problem of reconstructing coefficient signals from compressed sensing is an NP-hard problem [53]. In addition to sparsity, fully exploiting other prior knowledge has led to various image reconstruction strategies and methods, mainly including greedy algorithms [67, 68] and convex relaxation algorithms [11, 59, 82]. The sparse decomposition of signals under a redundant dictionary was the most classic greedy iterative algorithm proposed by Mallat et al. [49]. Because of its computational efficiency and ease of implementation, the algorithm has been widely used in image compressed sensing reconstruction [36, 69]. Subsequently, many scholars improved the above method and proposed an orthogonal matching pursuit (OMP) algorithm [28, 51, 68], which orthogonalizes the column vectors in the selected submatrix to accelerate convergence. In [46], the authors proposed a regularized orthogonal matching pursuit degree algorithm, which is better suited for image signal reconstruction. Thong et al. introduced the sparsity adaptive matching pursuit algorithm in [17]. Another common algorithm is to transform the nonconvex optimization problem in the compressed sensing reconstruction into a convex optimization problem. In [7], the authors proposed an iterative shrinkage threshold method, which was a commonly used convex optimization algorithm for compressed sensing reconstruction. Yin et al. [80] introduced the Bregman iterative regularization method into compressed sensing reconstruction and transformed the compressed sensing reconstruction problem into an unconstrained optimization problem. A method to perform image reconstruction using highly incomplete data by a special recursive filtering process was proposed [23]. In [20], an image reconstruction method based on an autoregressive model was proposed. Yu et al. used a mixed Gaussian model for training and reconstruction [81]. These methods exploit the statistical prior information of images in sparse representation and reconstruction models, enabling to obtain fast recovery and estimation of images. The image reconstruction algorithm based on nonlocal central clustering proposed by Dong et al. was also a classical reconstruction algorithm based on sparse characteristics [18, 19].

4.2.2.3 Deep Learning–Based Reconstruction/Generation

In recent years, the rapid development of deep learning [35, 56, 58] has brought new ideas and perspectives to image reconstruction/generation methods. In the work [47], Lohit et al. proposed a data-driven non-iterative algorithm to overcome the limitations of early iterative algorithms and used end-to-end deep CNNs to learn the measurement matrix and reconstruction algorithm in a single network. Moreover, using CNNs, the authors extracted hierarchical increasingly spatial features for reconstructing hyperspectral imagery, and the results were better than traditional methods [45]. In addition, in the work [61], the authors provided a new method for holographic image reconstruction based on deep learning. This algorithm can quickly eliminate artifacts related to twin images and self-interference using a single hologram intensity. The neural network framework proved to be more efficient than existing holographic phase recovery methods. Kelly et al. [42] used CNNs as a quasi-projection operator in the process of least squares minimization to encode high-level information in images. The proposed method improved the performance of image reconstruction in iterations. GANs [29] model the distribution of natural images by forcing samples that are indistinguishable from the original images. GANs have been widely used in image generation. Radford et al. [60] proposed a more stable GAN, whose network is capable of learning good representations of images for supervised learning and generative modeling. Reed et al. proposed a new model, named generative adversarial what-where network [60]. The images generated through this network can give instructions on what content to draw and where to place it. In the work [4], a new GAN algorithm, which can improve learning stability and eliminate mode collapse, was proposed and named Wasserstein GANs. The energy-based generative adversarial network model was proposed and trained to generate high-resolution images in the work [83]. Isola et al. [39] investigated conditional adversarial networks for image-to-image translation. This method was effective at synthesizing images from label maps. Furthermore, Zhu et al. [84] proposed an approach for learning to translate an image from a source domain to a target domain in the absence of the paired examples. Wang et al. [71] proposed multi-scale generator and discriminator architectures for high-resolution image generation.

4.2.3 Proposed Model

4.2.3.1 2D Sketch

The concept of "sketch" can be traced back to the seminal work of David Marr [5]. At the time, Marr did not provide a strict mathematical proof. Afterward, Songchun Zhu et al. [31] furthered Marr's work and provided a mathematical model of the sketch. A sketch map serves as a concise and structural representation of the image, capturing the positions of amplitude or intensity changes, as well as the shape and geometric structures of the image. In Zhu's paper, they divided images into the structure domain and texture domain according to the primal sketch. Then, a sparse coding model was used to represent image intensities of the structure domain, such as edges and ridges. The Markov random field model was used to summarize the intensities of the texture domain with some regular pattern. In [72], the ratio of average detector and cross-correlation-based (CC) detector were fused to detect edge-line features in SAR images, which are equally applicable to natural and panoramic images. We use the sketching model from [72] to compute the sketch map of the panoramic image. Simultaneously, the orientations of the edges are also computed in the sketching process. There are three parameters in the model of [72], including the coding length gain (CLG), high threshold (HT), and low threshold (LT) of NMS. The larger the CLG is, the sparser the sketch map is, i.e., the fewer the number of segments in the sketch map. It should be noted that the sketch map is obviously different from traditional edge detectors, such as the Canny detector. In fact, the filter of the Canny detector can be used as a subset of the filters in the sketching process. In addition, the contour map of the Canny detector consists of edge points. The sketch map consists of line segments with the orientation information and can be derived from the following steps [72]:

1. Compute the edge-line intensity map by adaptively fusing responses obtained from edge-line templates with different scales and orientations.
2. Extract the sketch curves using non-maximum suppression and a double-threshold-based connection method applied to the intensity map.
3. Extract the sketch lines from the sketch curses by the approximation method.
4. Generate the sketch map by evaluating the significance of each sketch line using a pair of paradoxical hypotheses and preserving the significant lines.

Figure 4.8 shows examples of sketch maps for panoramic images. Figure 4.8a presents the original panoramic images, while Fig. 4.8b displays the 2D sketch maps of the panoramic images with the thresholds $CLG = 15$, $HT = 1.5$, and $LT = 0.7$. Figure 4.8c and d show the edge maps of the Canny detector with $\sigma = 2$ and $\sigma = 4$, respectively. It is evident that the sketch maps provide clearer structural information compared to the edge maps produced by the Canny detector. For example, in the second row, the sketch map effectively represents the structures of the buildings, while the Canny detector just only indicates the positions of the edges. Moreover, the sketch line segments or edge points constitute only a small percentage of the total image pixels, as indicated in Fig. 4.8. Analogous to [38, 71], the sketch maps also utilize the boundary map as the input, but they also capture the shape and geometric structures. The code for the sketching model is available at http://web.xidian.edu.cn/fliu/lunwen.html.

The reasons for selecting the sketch map as the input are as follows: first, the sketch map provides a concise structural representation of the image, containing more details that can enhance the generated image quality. In contrast, the semantic map only contains the label information. In other words, the sketch map contains more structures than the semantic map. Certainly, both the sketch map and the semantic map can be used as the input to generate the panoramic image. Second, we

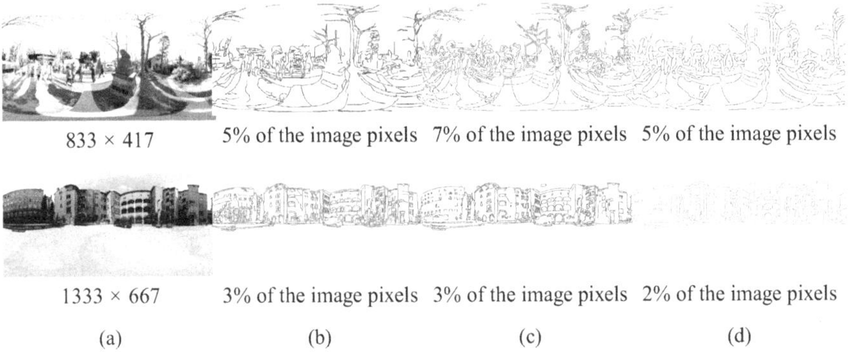

Fig. 4.8 Sketch maps of panoramic images. (**a**) shows the original panoramic images, (**b**) shows the corresponding sketch maps of the original images, (**c**) shows the edge contour using the Canny detector with $\sigma = 2$, and (**d**) shows the edge contour using the Canny detector with $\sigma = 4$

usually need complex computations to produce the semantic map in a supervised manner. However, the sketch map is computed by the predefined detectors in an unsupervised manner. Third, in the communication system, if the transmitted data are represented as a sketch map, it will greatly reduce the transmitted data volume of the communication. For example, for a panoramic image of size 583×1163, the sketch map contains 44,861 sketch points, accounting for approximately 7% of the total pixels.

4.2.3.2 Spherical Convolution

In recent years, CNNs have achieved remarkable results in speech recognition, visual object recognition, natural language processing, and other domains. CNNs excel in analyzing signals such as audios, texts, images, or videos due to their shift-invariance property, which confers powerful discriminative capabilities in computer vision. However, traditional CNN models lack rotational invariance and are unable to process data from multiple angles during training. Many modern applications, such as virtual reality and autopilot systems, utilize 360-degree cameras to capture spherical signals. Currently, spherical signals are often mapped to a two-dimensional space and processed using naive CNNs. However, these projections can introduce shape and size distortions, resulting in certain areas appearing larger or smaller than they are in the actual scene.

In [13], the authors pointed out that the equivariance is a good inductive bias in CNNs and proposed group equivariant CNNs. Afterward, they naturally extended the two-dimensional system to a three-dimensional system and proposed spherical CNNs [12], which are particularly effective for recognizing 3D shapes and performing various prediction tasks. In particular, $S2$ is a two-dimensional manifold in the spherical space and is defined as the set of points $x \in \mathbb{R}^3$ with a norm of 1.

The output feature map is computed as the inner product between the input feature map and the filter that has been rotated by R. According to [12], for spherical signals u and filter ψ, spherical correlation is defined as:

$$[\psi * u](R) = \langle L_{R\psi}, u \rangle = \int_{s^2} \sum_{k=1}^{K} \psi_k(R^{-1}x)u_k(x)dx \tag{4.5}$$

where $\langle \cdot \rangle$ denotes the inner product and dx represents the standard rotation-invariant integration measure on the sphere.

Similarly, the rotation group $SO(3)$ is a three-dimensional manifold representing the set of rotations. Similar to $S2$, the rotation group correlation is defined as:

$$[\psi * u](R) = \langle L_{R\psi}, u \rangle = \int_{SO(3)} \sum_{k=1}^{K} \psi_k(R^{-1}Q)u_k(Q)dQ \tag{4.6}$$

where dQ represents the invariant integration measure on $SO(3)$.

To solve the above equations, Cohen et al. [12] employed the fast Fourier Transform (FFT) to compute the correlations. The computational complexity of spherical convolution is $O(n \log n)$, dependent on the FFT. The spherical convolution can be computed using the following steps:

- Compute the FFT coefficients for the spherical signals u and the filters ψ, separately, resulting in coefficients $\widehat{u}$ and $\widehat{\psi}$.
- Compute the product of the FFT coefficients $\widehat{u}$ and $\widehat{\psi}$.
- Perform the inverse FFT on the product to obtain the output feature map.

4.2.3.3 Panoramic Image Generation Model

Panoramic images present significant challenges in terms of their generation due to their high resolutions Previously, DNN-based methods mainly focused on generating thumbnail images, such as the MNIST dataset with a size of 28 × 28, the CIFAR dataset with a size of 32 × 32 [64], and so on. Recent studies have begun exploring DNN-based methods for generating generate the high-resolution images [71, 84]. In our proposed approach, we mainly improve the pix2pixHD framework [71] by using the spherical convolution and 2D sketch. Under the GAN framework, our approach includes the generator G and the discriminator D. For our task, the generator is used to translate the 2D sketch into realistic-looking panoramic images, while the discriminator is used to distinguish the generated images from real ones. After the adversarial learning, a sketch map is translated into a spherical image.

Generator The generator's objective is to synthesize a spherical panoramic image from the 2D sketch. At the same time, we hope that ensuring the generated image is close enough to the original image. We design the spherical generator network to generate the spherical image. By using the spherical convolution, it can reduce the distortion due to the planar space projection. The training data comprise a set of pairs of images $\{(s_i, x_i)\}$, where s_i represents the 2D sketch and x_i represents the original panoramic image. To improve the fidelity and perceptual quality of the panoramic image, we consider three aspects in the generator design. First, the generator G generates images that are difficult for discriminator D to distinguish from real images. We use the loss of least squares GANs (LSGANs) to describe this constraint:

$$\mathcal{L}_{GAN} = \mathbb{E}_{s\sim p_{data}(s)}[(D(s, G(s)) - 1)^2] \tag{4.7}$$

Second, we use feature matching loss to ensure stable training. Similar to [71], we extract features from the discriminator at different layers for both the generated and real images. On one hand, the feature makes the training process stable. On the other hand, it acts as a perceptual loss to improve the quality of the generated image. Assume that N_L is the total number of layers and F_l^{VGG} represents the feature extracted from the lth layer of the discriminator. The feature matching loss is defined as:

$$\mathcal{L}_{FM} = \mathbb{E}_{(s,x)} \sum_{l=1}^{L_1} \frac{1}{N_l} \left[\left\| F_l^{\text{VGG}}(s, x) - F_D^i(s, G(s)) \right\|_1 \right] \tag{4.8}$$

where N_l is the number of the elements in the lth layer.

Third, the perceptual quality is an important evaluation index for image generation, especially in DNN-based methods. Achieving plausible QoE in the generated image is the ultimate goal. Therefore, we should emphasize this ability when constructing the model. By optimizing the similarity in the feature space, it can improve the semantic similarity between the generated image and the real image. This loss has been used in several works to improve the perceptual quality [22, 44, 71]. In particular, for the generated image $G(s)$ and the real image x, their features are extracted from the VGG network. The corresponding perceptual loss is computed as follows:

$$\mathcal{L}_{perceptual} = \sum_{l}^{N_L^{\text{VGG}}} \frac{1}{M_i} \left[\left\| F_{VGG}^i(s, x) - F_{VGG}^i(s, G(s)) \right\|_1 \right] \tag{4.9}$$

where N_L^{VGG} is the total number of layers in the VGG network. M_l is the number of elements in the lth layer. F_{VGG}^i represents the feature extracted from the ith layer of the VGG network.

Based on the above, we combine the three losses as the objective function of the generator:

$$\mathcal{L}_{final} = \mathcal{L}_{GAN} + \mathcal{L}_{FM} + \mathcal{L}_{perceptual} \tag{4.10}$$

Discriminator The discriminator aims to differentiate between the generated and real images. For real images, the discriminator wants its output to be true. For generated images $G(s)$, the discriminator wants its output to be false. The loss of the discriminator is written as:

$$\mathcal{L}_G = \frac{1}{2}\mathbb{E}_{(s,x)\sim p_{data}(s,x)}[(D(s,x) - 1)^2] + \frac{1}{2}\mathbb{E}_{s\sim p_{data}(s)}[(D(s, G(s)))^2] \tag{4.11}$$

Additionally, we employ multi-scale discriminators D_1, D_2, and D_3 as commonly used in image generation [71]. The coarse-to-fine paradigm improves the quality of the generated image. At the coarse scale, it captures global information and improves consistency of the generated image with the large receptive field, while at the fine scale, it captures the information at the local view and preserves the details, such as edges and lines. Moreover, the multi-scale method can reduce the burden on the network and make the network easier for training.

4.2.3.4 Network Architecture

We enhance the network architecture introduced in [71], while retaining the same settings for common hyperparameters. The network architecture of our proposed approach is illustrated in Fig. 4.9, representing the forward process of the data flow. Starting from the left, the input is the 2D sketch map. Through the spherical generator network, discriminator network, and VGG network, five losses are computed. In the backward process, we optimize these losses to train the model.

The generator adopts an "encoder-decoder" structure. The input of the generator is the 2D sketch, which needs to be projected onto the sphere using the method described in [12] for preprocessing. We follow the definition used in [12] and [71]. Assume that spherical convolution $S2$, normalization, and activation function is a tuple and defined as $S2-k-B_w$, where k is the number of the input/output channel and B_w is the input/output bandwidth. R_{S2-k-B_w} is a residual block that consists of two $S2-k-B_w$ tuples. Here, we choose ReLU as the activation function and Instance3d for normalization. The generator starts with a naive convolutional layer with a kernel size of 7×7. The second block is composed of three downsampling layers, followed by five residual blocks. Symmetrically, three upsampling layers are included, and a convolutional layer is used to generate the three-dimensional output. Following the notations in [71], the spherical generator network contains $c7s1-26$, $d208$, $d104$, $d52$, $(R_{S2-208-32})\times 5$, $u52$, $u104$, $u208$, and $c7s1-3$. Here, $R_{S2-208-32}$ denotes $S2$ convolution with the 3×3 kernel, input/output channel 208, and input/output bandwidth 32. The generator structure is shown in Fig. 4.10. We do not use $SO(3)$ convolution as it requires a huge amount of memory and computation. Moreover, the rotation along the γ dimension is not included in our experiments (dataset). Using $S2$ convolution helps to reduce memory usage.

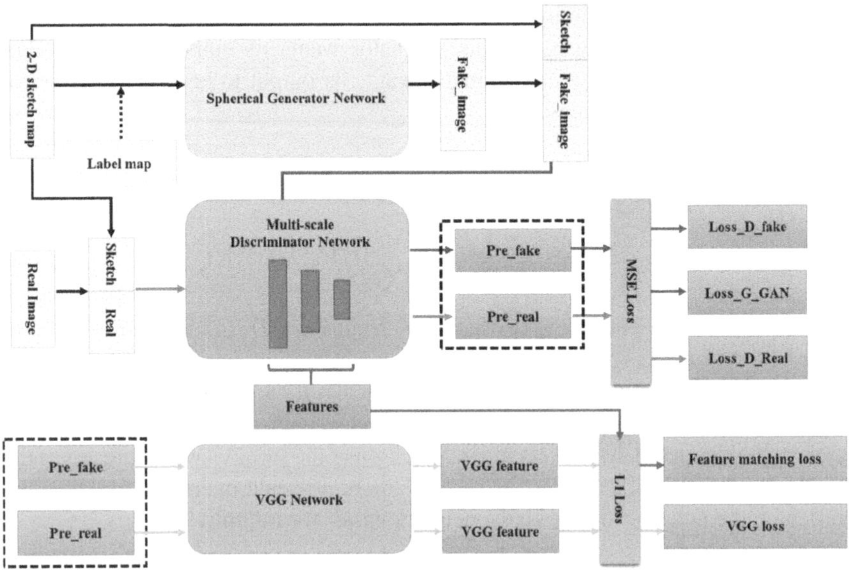

Fig. 4.9 Structure of the proposed model, illustrating the flow of data from input to output and the forward process of the proposed model. Through the generator and the discriminator, five losses are obtained. By optimizing the network using these loss functions, the desired model is obtained

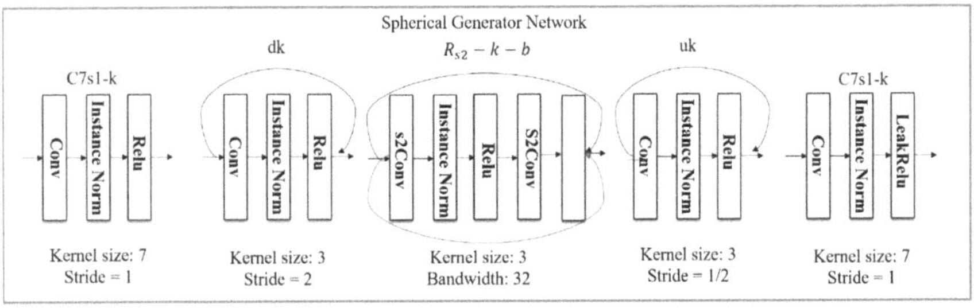

Fig. 4.10 Structure of the generator. The output of each $S2$ convolution is averaged along the γ dimension to transform the feature map back to 2D. The black arrow represents the presence of several corresponding tuples in the network, while the red arrow represents the residual block

For the discriminator, we follow the same network architecture as in [71]. We do not employ spherical convolution in the discriminator for two reasons. First, the primary purpose of the discriminator is to differentiate between generated and real images, which is independent of whether the image is two-dimensional or spherical. Second, using spherical convolution would increase the network's storage and computational burden. Therefore, we only utilize spherical convolution in the generator to generate the spherical image, thereby avoiding image distortion and enhancing the generated image's quality.

4.2.4 *Result and Analysis*

This section presents the experimental results conducted in two parts to verify the effectiveness of the proposed approach. In the first part, we only use the sketch map as input to synthesize the panoramic image. The experiments are mainly used to prove that textures can be synthesized where no another input information is provided. In the second part, we improve the synthesis quality by incorporating the semantic label map. These experiments confirm that providing more information leads to improved quality. In addition, it should be noted that we train different models for different datasets due to their differing distributions. Training both datasets with one model may lead to interference. The code for our approach is at https://github.com/hancy16/SGAN.

4.2.4.1 Results on AOI Dataset

We develop the original panoramic image dataset (AOI dataset) [77] for panoramic image generation. We construct a corresponding sketch map dataset using the sketching model described in Sect. 4.2.3.1. Each pair of an original panoramic image and its sketch map form a new dataset for panoramic image generation. The dataset includes 556 panoramic images and 556 sketch maps. These images are collected from different scenarios, classified into Cityscapes, Hall_exhibition, Human_party, Human_tour, Indoor_rooms, and Natural_landscapes. The image resolutions range from 299 × 597 to 1333 × 667. The sketch map is a binary image, where sketch lines are denoted as 0 and the rest as 255. However, other value definitions, such as 0 and 1, can also be used without affecting the training of the proposed model. We divide the dataset into a training dataset (500 pairs) and a testing dataset (56 pairs). To the best of our knowledge, this is the first dataset utilizing a sketch map for panoramic image generation. Examples from the AOI dataset are shown in Fig. 4.11.

Fig. 4.11 Some real images from AOI dataset

Fig. 4.12 Visual examples of the images generated on the AOI dataset. The first and third columns display the sketch maps, while the second and fourth columns show the synthesized images produced by the proposed approach

Experimental Setup With the dataset, we train the model on the training dataset from scratch, which contains 500 panoramic images and sketch maps. The testing set, comprising 56 pairs, is used for evaluation. Due to the substantial memory requirements of spherical convolution, the inputs are resized to 512×512 pixels. All convolutional kernels are initialized with a Gaussian distribution (mean = 0, and standard deviation = 0.02). The weights of InstanceNorm3d are initialized with a Gaussian distribution (mean = 1, and standard deviation = 0.02). We use the Adam solver with a learning rate of 0.0002 as the optimizer. The batch size is set to 1. The proposed model is trained on 4 TITAN XP GPUs, and training is completed after 100 epochs.

Result The generation results of panoramic images using only sparse sketch information are shown in Fig. 4.12. We resize the generated images to their original sizes for visualization. In the AOI dataset, the ratio of sketch lines to total pixels is approximately 7%. The adversarial learning process progressively recovers textures and details from the very sparse sketch map. In [38, 56, 84], they also use the sparse contours to generate the image. However, their databases typically consist of a single type of image, such as VGG faces or Stanford Dogs, with simple scene content. For different datasets, they train different models. Panoramic images with high resolution contain rich textures and details, posing significant challenges for panoramic image generation. From Fig. 4.12, we can see that the proposed approach produces high quality reconstruction. In particular, the edge portion of the synthesized image is clear. For example, in the third row of Fig. 4.12, the panoramic images contain complex scenery, buildings, and some people. The proposed model generates rich and complicated textures and details. From the visual quality, the

Fig. 4.13 Visual examples of the images generated on the AOI dataset. The first and third columns display the sketch maps, while the second and fourth columns show the synthesized images produced by the proposed approach

generated images are plausible and have a realistic-looking appearance. These images can also be used in subsequent processing, such as semantic segmentation, object recognition, and other visual tasks.

Comparison We compare the proposed method with the baseline Pix2pixHD model [71]. The corresponding experimental results are shown in Fig. 4.13. To ensure a fair comparison, we retrain the baseline model using the same settings and details. The Pix2pixHD model is an improved version of the Pix2pix model [39]. Using the coarse-to-fine generator, multi-scale discriminator and feature matching loss, the Pix2pixHD model improves the generated image quality and is suitable for generating high-resolution images. However, as shown in the third column of Fig. 4.13, colors appear strange. For example, in the first image, the colors of the sky and ground are mixed. Moreover, the generated image structures are implausible, especially in the fourth image. By using the proposed approach, the visual quality is improved and the textures are recovered clearly. In the proposed method, we plug the spherical convolution into the generator to reduce the distortions. With the proposed model, we implement the panoramic image generation using only the sparse sketch maps. From Fig. 4.13, we can see that the visual quality of the proposed method is better than that of the baseline model. This demonstrates the necessity of spherical convolution for spherical signals/images and the effectiveness of perceptual loss in high-level image generation.

4.2.4.2 Results on SYNTHIA Dataset

SYNTHIA Dataset The SYNTHIA dataset [62] is generated from a visual world of urban scenarios with semantic labels available. Such class annotations are shown to be essential for high-quality image generation, and they require minimal storage capacity, e.g., an average of 0.036 bpp on average for downscaled Cityscapes [14] images, as reported in [1]. Moreover, the dataset provides multiview images, making it an ideal choice for the semantic panoramic dataset. Therefore, we conduct the panoramic image generation task on the SYNTHIA dataset and include its semantic label maps.

Spherical Projection The SYNTHIA dataset comprises several video sequences, with each frame consisting of eight views. Since the original data consist of regular images, we generate panoramic images by projecting them onto a sphere. Specifically, for each frame, we utilize the four images captured by the left stereo-camera and transform them into a single panoramic image. To accomplish this, we first select a reference spherical coordinate system R_0 and assume constant distances between the images and the origin, denoted as f. For an image $I(x, y)$, we obtain its coordinates in R_0 using the rotation transformation:

$$\begin{bmatrix} u \\ v \\ w \end{bmatrix} = \begin{bmatrix} \cos\beta & 0 & \sin\beta \\ 0 & 1 & 0 \\ -\sin\beta & 0 & \cos\beta \end{bmatrix} \times \begin{bmatrix} 1 & 0 & 0 \\ 0 & \cos\alpha & -\sin\alpha \\ 0 & \sin\alpha & \cos\alpha \end{bmatrix} \times \begin{bmatrix} x - \frac{W}{2} \\ y - \frac{H}{2} \\ f \end{bmatrix} \tag{4.12}$$

where $(W_{\text{image}}, H_{\text{image}})$ denotes the width and height of I, (α_r, β_r) denotes the rotation relative to R_0, and f_{cam} actually denotes the camera's focus. In our implementation, f_{cam} is calculated as $W_{\text{image}}/[2\tan(\frac{\tau_{\text{cam}}}{2})]$, where τ_{cam} is the field angle of the camera. We project the images onto a sphere with a radius of f_{cam}. The spherical coordinates can then be computed as:

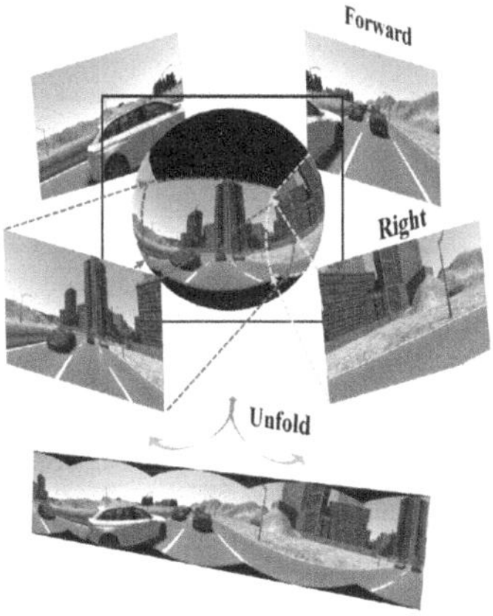

Fig. 4.14 Illustration of the spherical projection, where the four views are seamlessly stitched

Table 4.1 Parameter setting for spherical projection

View	α	β	τ
Front	0	0	100
Left	0	90	100
Back	0	180	100
Right	0	270	100

Fig. 4.15 Some spherical images from the SYNTHIA dataset

$$\theta = \arctan\left(\frac{u}{w}\right)$$
$$\phi = \arctan\left(\frac{v}{\sqrt{u^2 + w^2}}\right) \quad (4.13)$$

After projecting all the images, we average the intensities in overlapping regions and crop the resulting 2D sphere, as the top and bottom views are absent. The entire process is illustrated in Fig. 4.14. The parameter settings are provided in Table 4.1. During the processing, we downsample the original images by a factor of 0.5, and the final panoramic images have a size of 1687×335 pixels. Examples from the dataset are shown in Fig. 4.15. The datasest is available at https://drive.google.com/drive/u/0/folders/10BF1hZQnndj-Yi-2vX7gGwww1f6i8jP9.

Experimental Setup We utilize the SYNTHIA-SEQ5-05-SUMMER dataset, generating a total of 787 panoramic images. Out of these, 579 images are randomly selected as the training set, and the remaining images form the test set. During

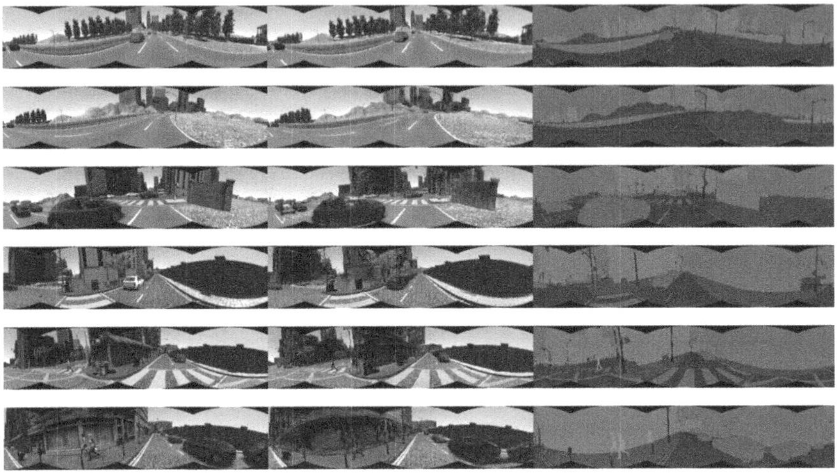

Fig. 4.16 Visual examples of the images generated on the SYNTHIA dataset. From left to right: input sketch map, synthesized image, and input label map

training, all images are resized to 512 × 512 pixels. The networks are trained from scratch for 100 epochs, requiring approximately 32h on 4 TITAN XP GPUs.

Results In Fig. 4.16, we present several random samples of the generated images. The first column is the sketch map, the second column displays the generated image, and the third column shows the input label map. It can be observed that the generated images are often realistic and plausible. In particular, the last row demonstrates the recovery of small objects, such as people and pillars, from sparse inputs. Homogeneous regions, such as road and sky, are also synthesized with high quality. The use of the sketch map preserves more details, and the label map enhances the consistency of specified label. However, due to resizing all inputs to 512 × 512 pixels, the resolution of the generated images is slightly lower than that of the original images. For visualization purposes, we resize the generated images to the original size.

Comparison Similar to the AOI dataset, we compare the proposed method with the baseline Pix2pixHD model. The corresponding experimental results are shown in Fig. 4.17. The first column depicts the generated images using our approach, while the second column shows the generated images using the Pix2pixHD model. These results appear comparable, but the edge portions in our proposed approach are clearer, especially the building edges in the third row. This demonstrates that the sketch map improves the quality of the edge regions. Furthermore, in the first and second rows, the shadows in the generated images from the baseline model appear strange and discontinuous, while our method avoids this phenomenon by employing spherical convolution, which is more suitable for generating spherical structures. These experimental results validate the effectiveness of the proposed approach.

Fig. 4.17 Visual examples of the generated images using our approach and a baseline model on the SYNTHIA dataset. From left to right: our approach, pix2pixHD baseline model

4.2.4.3 Quantitative Evaluations

Objective Evaluation To further evaluate the performance of the proposed approach, we adopt the inception score (IS) [63, 78] and the Frechet inception distance (FID) [34, 78] as objective measures to quantify the performance of the generative models. The IS score reflects the plausibility and diversity of the generated images, where a higher score indicates better results. The FID score measures the distance between the generated images and the real images, with a smaller score indicating closer resemblance to real images. The numerical indexes for the AOI dataset and SYNTHIA dataset are presented in Tables 4.2 and 4.3, respectively. For the AOI dataset, we compute the IS score and FID score using 56 testing images, and the results are shown in Table 4.2. The IS mean represents the average score across all test images, while the IS std denotes the standard deviation. From Table 4.2, we observe that the IS mean of the Pix2pixHD model is slightly higher than that of the proposed approach. We speculate that this phenomenon is caused by the presence of checkerboard artifacts in the Pix2pixHD model. Furthermore, it should be noted that the IS model is trained on the ImageNet dataset and may not accurately reflect the capabilities of our model. Nevertheless, we still provide the IS scores for reference. The FID scores of the proposed approach are 24 points higher than those of the baseline model. For the SYNTHIA dataset, we compute the IS score and FID score using 200 testing images, and the results are presented in Table 4.3. We can observe that both the IS scores and FID scores of the proposed approach outperform those of the baseline model.

Subjective Evaluation In addition, we provide quantitative evaluations of the generated images from a human perspective. We invited 30 human evaluators to compare the generated images from the proposed approach and the Pix2pixHD

Table 4.2 Quantitative results on AOI dataset. We report IS scores (higher is better) and FID scores (lower is better) for test images

AOI dataset			
Method	IS mean	IS std	FID
Ours	1.718	0.208	208.743
Pix2pixHD	1.727	0.380	232.375

Table 4.3 Quantitative results on SYNTHIA dataset. We report IS scores (higher is better) and FID scores (lower is better) for test images

SYNTHIA dataset			
Method	IS mean	IS std	FID
Ours	1.965	0.330	80.735
Pix2pixHD	1.895	0.367	84.845

Table 4.4 Human evaluation results. We report the preference percentages of our results and the pix2pixHD results

Preference of our results	Preference of pix2pixHD results
66%	34%

method. For each sample, the evaluators were presented with two generated images and were asked to select the better one. We prepared 30 pairs of generated images from the AOI dataset and SYNTHIA dataset, with each pair consisting of images generated by the proposed approach and the baseline model. The corresponding results are displayed in Table 4.4. It can be observed that the baseline model only achieves a 34% preference rate, while our model achieves a 66% preference rate. From a human viewpoint, our model outperforms the baseline model by 30%. These results indicate that the generated images produced by our proposed approach exhibit higher quality.

4.2.4.4 Ablation Experiments

To analyze the effects of different components, we report the IS scores and FID scores on the AOI dataset for several ablations of the proposed model. For clarity, we define the full model as $GAN + SC + L_{FM} + L_{VGG}$.

Spherical Convolution Spherical convolution is primarily employed to capture rotational invariance and multiple angles during the training process. It helps reduce distortion caused by 2D planar mapping. In fact, the difference between the proposed approach and the baseline model is the spherical convolution. The impact of spherical convolution is shown in Tables 4.2 and 4.3.

Feature Matching Loss The feature matching loss is used to align the intermediate representations of real and generated images. To observe its effect, we retrain a model without L_{FM} denoted as $GAN + SC + L_{VGG}$.

Table 4.5 Ablation experiments on AOI dataset. We report IS scores (higher is better) and FID scores (lower is better) for generated images

AOI dataset			
Methods	IS mean	IS std	FID
$GAN + SC + L_{FM} + L_{VGG}$ (Ours)	1.718	0.208	208.743
$GAN + L_{FM} + L_{VGG}$ (Pix2pixHD)	1.727	0.380	232.375
$GAN + SC + L_{VGG}$	1.867	0.365	265.549
$GAN + SC + L_{FM}$	1.686	0.322	254.708

VGG Loss VGG loss, a form of perceptual loss, computes the similarity in the representation space. We also retrain a model without L_{VGG}, denoted as $GAN + SC + L_{FM}$.

The corresponding experimental results are displayed in Table 4.5. From Table 4.5, we observe that without spherical convolution, the IS scores are almost identical to those of the proposed approach. However, the FID scores of the proposed approach are higher compared to the $GAN + L_{FM} + L_{VGG}$ model. This indicates that spherical convolution can reduce distortion in panoramic image generation. Removing the feature matching loss leads to higher IS scores, possibly due to the reduction of checkerboard artifacts. However, the FID scores increase significantly, highlighting the significant impact of feature matching loss on image generation. We also provide the performance without VGG loss, which results in lower IS scores and higher FID scores compared to the proposed approach. This demonstrates that incorporating perceptual loss enhances the results.

4.2.5 Conclusion

This chapter has presented an approach for panoramic image generation. Within the framework of GANs, we have designed a generator with spherical convolution to preserve the fidelity of panoramic images. Moreover, we have demonstrated the ability to generate high-resolution panoramic images using only sparse sketch maps as input. This advancement is particularly significant in terms of communication, as it enables the transmission of a large amount of data with minimal bandwidth. In addition, we have combined GAN loss, feature matching loss, and VGG loss in the objective function. The experimental results have shown that the proposed approach is capable of generating high-resolution panoramic images that exhibit a realistic and plausible quality. This approach holds potential for various applications, such as communication systems, where transmitted data can be compressed into sketch maps, resulting in significant bandwidth savings.

In the future, we aim to address two key challenges. First, panoramic images are generated based on high-level semantics and structures, thereby achieving high visual quality. However, pixel-wise quality is also important in certain applications. Therefore, it is necessary to strike a balance between subjective and objective qualities by integrating combine high-level semantics with low-level features. Second, the requirement for square-shaped input and low-resolution images, as well as the memory-intensive nature of spherical convolution, poses limitations to the model's capability. Additionally, optimizing spherical convolution and enhancing the model's generality are also among our future research directions.

References

1. Agustsson, E., Tschannen, M., Mentzer, F., Timofte, R., Van Gool, L.: Generative adversarial networks for extreme learned image compression. arXiv preprint arXiv:1804.02958 (2018)
2. Antonini, M., Barlaud, M., Mathieu, P., Daubechies, I.: Image coding using wavelet transform. IEEE Trans. Image Process. **1**(2), 205–220 (1992)
3. Arbelaez, P., Maire, M., Fowlkes, C., Malik, J.: Contour detection and hierarchical image segmentation. IEEE Trans. Pattern Anal. Mach. Intell. **33**(5), 898–916 (2010)
4. Arjovsky, M., Chintala, S., Bottou, L.: Wasserstein gan. In: International Conference on Machine Learning (2017)
5. Barlow, H.B.: Vision: A computational investigation into the human representation and processing of visual information: David Marr. San Francisco: W. H. Freeman, 1982. pp. xvi + 397. J. Math. Psychol. **27**(1), 107–110 (1983)
6. Bhattacharya, C., Mahapatra, P.R.: A discrete wavelet transform approach to multiresolution complex sar image generation. IEEE Geosci. Remote Sens. Lett. **4**(3), 420 (2007)
7. Blumensath, T., Davies, M.E.: Iterative hard thresholding for compressed sensing. Appl. Comput. Harmon. Anal. **27**(3), 265–274 (2009)
8. Brisaboa, N.R., Ladra, S., Navarro, G.: K2-trees for compact web graph representation. In: Proceedings of the 16th International Symposium on String Processing and Information Retrieval (SPIRE '09), pp. 18–30. Springer, Berlin (2009). https://doi.org/10.1007/978-3-642-03784-9_3
9. Canny, J.: A computational approach to edge detection. IEEE Trans. Pattern Anal. Mach. Intell. **8**(6), 679–698 (1986)
10. Claude, F., Navarro, G.: Fast and compact web graph representations. ACM Trans. Web **4**(4), 1–31 (2010). https://doi.org/10.1145/1841909.1841913
11. Cohen, A., Dahmen, W., Devore, R.: Compressed sensing and best *k*-term approximation. J. Am. Math. Soc. **22**(1), 211–231 (2009)
12. Cohen, T.S., Geiger, M., Koehler, J., Welling, M.: Spherical CNNS (2018). https://arxiv.org/abs/1801.10130
13. Cohen, T.S., Welling, M.: Group equivariant convolutional networks (2016). http://arxiv.org/abs/1602.07576
14. Cordts, M., Omran, M., Ramos, S., Rehfeld, T., Enzweiler, M., Benenson, R., Franke, U., Roth, S., Schiele, B.: The cityscapes dataset for semantic urban scene understanding. In Proceedings of the IEEE Conference on Computer Vision and Pattern Recognition (CVPR), pp. 3213–3223 (2016)
15. Dekel, T., Gan, C., Krishnan, D., Liu, C., Freeman, W.T.: Smart, sparse contours to represent and edit images. CoRR **abs/1712.08232** (2017). http://arxiv.org/abs/1712.08232
16. Deng, J., Dong, W., Socher, R., Li, L.J., Li, K., Fei-Fei, L.: Imagenet: A large-scale hierarchical image database. In: Proceedings of the IEEE Conference on Computer Vision and Pattern Recognition, pp. 248–255 (2009)
17. Do, T.T., Gan, L., Nguyen, N., Tran, T.D.: Sparsity adaptive matching pursuit algorithm for practical compressed sensing. In: Conference on Signals, Systems & Computers (2010)
18. Dong, W., Lei, Z., Shi, G.: Centralized sparse representation for image restoration. In: IEEE International Conference on Computer Vision (2012)

19. Dong, W., Zhang, L., Shi, G., Li, X.: Nonlocally centralized sparse representation for image restoration. IEEE Trans. Image Process. **22**(4), 1620–1630 (2013)
20. Dong, W., Zhang, L., Shi, G., Wu, X.: Image deblurring and super-resolution by adaptive sparse domain selection and adaptive regularization. IEEE Trans. Image Process. **20**(7), 1838–1857 (2011)
21. Donoho, D.L.: Compressed sensing. IEEE Trans. Inf. Theory **52**(4), 1289–1306 (2006)
22. Dosovitskiy, A., Brox, T.: Generating images with perceptual similarity metrics based on deep networks. In: Lee, D.D., Sugiyama, M., Luxburg, U.V., Guyon, I., Garnett, R. (eds.) Advances in Neural Information Processing Systems, vol. 29, pp. 658–666. Curran Associates, Inc., Glasgow (2016). http://papers.nips.cc/paper/6158-generating-images-with-perceptual-similarity-metrics-based-on-deep-networks.pdf
23. Egiazarian, K., Foi, A., Katkovnik, V.: Compressed sensing image reconstruction via recursive spatially adaptive filtering. IEEE Int. Conf. Image Process **1**, 549–552 (2007)
24. Eitz, M., Haysy, J., Alexa, M.: How do humans sketch objects? ACM Trans. Graph. **31**(4CD), 1–10 (2012)
25. Elshoura, S.M., Megherbi, D.B.: Analysis of noise sensitivity and reconstruction accuracy of tchebichef moments. In: Southeastcon (2008)
26. Fei-Fei, L., Fergus, R., Perona, P.: Learning generative visual models from few training examples: an incremental bayesian approach tested on 101 object categories. In: Proceedings of the IEEE Conference on Computer Vision and Pattern Recognition Workshop, pp. 178–178. IEEE, New York (2004)
27. Figueiredo, M.A.T., Nowak, R.D., Wright, S.J.: Gradient projection for sparse reconstruction: application to compressed sensing and other inverse problems. IEEE J. Sel. Top. Sign. Proces. **1**(4), 586–597 (2007). https://doi.org/10.1109/JSTSP.2007.910281
28. Goklani, H.S., Sarvaiya, J.N., Fahad, A.M.: Image reconstruction using orthogonal matching pursuit (OMP) algorithm. In: International Conference on Emerging Technology Trends in Electronics (2015)
29. Goodfellow, I.J., Pouget-Abadie, J., Mirza, M., Xu, B., Warde-Farley, D., Ozair, S., Courville, A., Bengio, Y.: Generative adversarial networks. Adv. Neural Inf. Proces. Syst. **3**, 2672–2680 (2014)
30. Gunturk, B.K., Altunbasak, Y., Mersereau, R.M.: Super-resolution reconstruction of compressed video using transform-domain statistics. IEEE Trans. Image Process. **13**(1), 33–43 (2004)
31. Guo, C.E., Zhu, S.C., Ying, N.W.: Primal sketch: Integrating structure and texture. Comput. Vis. Image Underst. **106**(1), 5–19 (2007)
32. Ha, D., Eck, D.: A neural representation of sketch drawings. In: International Conference on Learning Representations, pp. 1–15 (2018)
33. He, J., Zhang, S., Yang, M., Shan, Y., Huang, T.: Bi-directional cascade network for perceptual edge detection. In: Proceedings of the IEEE/CVF Conference on Computer Vision and Pattern Recognition, pp. 3828–3837 (2019)
34. Heusel, M., Ramsauer, H., Unterthiner, T., Nessler, B., Hochreiter, S.: Gans trained by a two time-scale update rule converge to a local nash equilibrium. In: Guyon, I., Luxburg, U.V., Bengio, S., Wallach, H., Fergus, R., Vishwanathan, S., Garnett, R. (eds.) Advances in Neural Information Processing Systems, vol. 30, pp. 6626–6637. Curran Associates, Inc., Glasgow (2017)
35. Hinton, G.E., Osindero, S., Teh, Y.W.: A fast learning algorithm for deep belief nets. Neural Comput. **18**(7), 1527–1554 (2014)
36. Hong, F., Yang, H.: A new compressed sensing-based matching pursuit algorithm for image reconstruction. In: International Congress on Image & Signal Processing (2013)
37. Hong, Y.L., Eom, I.K., Kim, Y.S.: Adaptive reconstruction of lost block using difference of dc in dct domain. In: International Conference on Signal Processing (2004)
38. Isola, P., Zhu, J., Zhou, T., Efros, A.A.: Image-to-image translation with conditional adversarial networks. In: 2017 IEEE Conference on Computer Vision and Pattern Recognition (CVPR), pp. 5967–5976 (2017). https://doi.org/10.1109/CVPR.2017.632

39. Isola, P., Zhu, J.Y., Zhou, T., Efros, A.A.: Image-to-image translation with conditional adversarial networks. In: IEEE Conference on Computer Vision and Pattern Recognition (2016)
40. Johnson, J., Gupta, A., Fei-Fei, L.: Image generation from scene graphs. In: Proceedings of the IEEE/CVF Conference on Computer Vision and Pattern Recognition, pp. 1219–1228 (2018)
41. Kawanishi, T., Yamazawa, K., Iwasa, H., Takemura, H., Yokoya, N.: Generation of high-resolution stereo panoramic images by omnidirectional imaging sensor using hexagonal pyramidal mirrors. In: Proceedings. Fourteenth International Conference on Pattern Recognition (Cat. No.98EX170), vol. 1, pp. 485–489 (1998). https://doi.org/10.1109/ICPR.1998.711187
42. Kelly, B., Matthews, T.P., Anastasio, M.A.: Deep learning-guided image reconstruction from incomplete data. In: Conference on Neural Information Processing Systems (2017)
43. Konishi, S., Song, C.Z., Yuille, A.L., Coughlan, J.: Fundamental bounds on edge detection: an information theoretic evaluation of different edge cues. In: IEEE Computer Society, p. 1573 (1999)
44. Ledig, C., Theis, L., Huszár, F., Caballero, J., Cunningham, A., Acosta, A., Aitken, A., Tejani, A., Totz, J., Wang, Z., Shi, W.: Photo-realistic single image super-resolution using a generative adversarial network. In: 2017 IEEE Conference on Computer Vision and Pattern Recognition (CVPR), pp. 105–114 (2017). https://doi.org/10.1109/CVPR.2017.19
45. Li, Y., Xie, W., Li, H.: Hyperspectral image reconstruction by deep convolutional neural network for classification. Pattern Recogn. **63**, 371–383 (2017)
46. Lin, Z.: Image recovery based on compressive sensing and curvelet transform via romp. In: 9th International Conference on Fuzzy Systems and Knowledge Discovery, pp. 1812–1815 (2012)
47. Lohit, S., Kulkarni, K., Kerviche, R., Turaga, P., Ashok, A.: Convolutional neural networks for non-iterative reconstruction of compressively sensed images. IEEE Trans. Comput. Imaging **PP**(99), 1–1 (2017)
48. Lu, J., Yang, Y., Liu, R., Kang, S.B., Yu, J.: 2d-to-stereo panorama conversion using gan and concentric mosaics. IEEE Access **7**, 23187–23196 (2019). https://doi.org/10.1109/ACCESS.2019.2899221
49. Mallat, S.G., Zhifeng, Z.: Matching pursuits with time-frequency dictionaries. IEEE Trans. Signal Process. **41**(12), 3397–3415 (1993)
50. Marr, D.: Vision: A Computational Investigation into the Human Representation and Processing of Visual Information, pp. xvii+397. WH Freeman and Company, New York (1982)
51. Meenakshi, Budhiraja, S.: Image reconstruction using modified orthogonal matching pursuit and compressive sensing. In: International Conference on Computing (2015)
52. Mentzer, F., Toderici, G.D., Tschannen, M., Agustsson, E.: High-fidelity generative image compression. Adv. Neural Inf. Proces. Syst. **33**, 11913–11924 (2020)
53. Ming, Y., Yi, L.: Model selection and estimation in regression with grouped variables. J. R. Stat. Soc. **68**(1), 49–67 (2006)
54. Mukundan, R., Ong, S.H., Lee, P.A.: Image analysis by tchebichef moments. IEEE Trans. Image Process. **10**(9), 1357–1364 (2001)
55. Newell, A., Yang, K., Deng, J.: Stacked hourglass networks for human pose estimation. In: European Conference on Computer Vision, pp. 483–499 (2016)
56. van den Oord, A., Kalchbrenner, N., Espeholt, L., kavukcuoglu, k., Vinyals, O., Graves, A.: Conditional image generation with pixelcnn decoders. In: Lee, D.D., Sugiyama, M., Luxburg, U.V., Guyon, I., Garnett, R. (eds.) Advances in Neural Information Processing Systems 29, pp. 4790–4798. Curran Associates, Inc., Glasgow (2016). http://papers.nips.cc/paper/6527-conditional-image-generation-with-pixelcnn-decoders.pdf
57. Orhei, C., Bogdan, V., Bonchis, C.: Dilated filters for edge detection algorithms. Appl. Sci. **11**(22), 10716 (2021)
58. O'Shea, T.J., Roy, T., Clancy, T.C.: Over-the-air deep learning based radio signal classification. IEEE J. Sel. Top. Sign. Proces. **12**(1), 168–179 (2018). https://doi.org/10.1109/JSTSP.2018.2797022

59. Osher, S., Mao, Y., Dong, B., Yin, W.: Fast linearized bregman iteration for compressive sensing and sparse denoising. Math. Comput. **8**(1), 93–111 (2010)
60. Reed, S., Akata, Z., Mohan, S., Tenka, S., Lee, H.: Learning What and Where to Draw. New Republic, New York (2016)
61. Rivenson, Y., Zhang, Y., Güaydin, H., Da, T., Ozcan, A.: Phase recovery and holographic image reconstruction using deep learning in neural networks. Light Sci. Appl. **7**(2), 17141–17141. (2017)
62. Ros, G., Sellart, L., Materzynska, J., Vazquez, D., Lopez, A.M.: The synthia dataset: a large collection of synthetic images for semantic segmentation of urban scenes. In: The IEEE Conference on Computer Vision and Pattern Recognition (CVPR) (2016)
63. Salimans, T., Goodfellow, I., Zaremba, W., Cheung, V., Radford, A., Chen, X., Chen, X.: Improved techniques for training gans. In: Lee, D.D., Sugiyama, M., Luxburg, U.V., Guyon, I., Garnett, R. (eds.) Advances in Neural Information Processing Systems 29, pp. 2234–2242. Curran Associates, Inc., Glasgow (2016). http://papers.nips.cc/paper/6125-improved-techniques-for-training-gans.pdf
64. Santurkar, S., Budden, D.M., Shavit, N.: Generative compression. CoRR **abs/1703.01467** (2017). http://arxiv.org/abs/1703.01467
65. Starck, J.L., Bijaoui, A., Lopez, B., Perrier, C.: Image reconstruction by the wavelet transform applied to aperture synthesis. Astron. Astrophys. **283**(283), 349–360 (1994)
66. Su, Z., Liu, W., Yu, Z., Hu, D., Liu, L.: Pixel difference networks for efficient edge detection. In: Proceedings of the IEEE International Conference on Computer Vision, pp. 5097–5107 (2021)
67. Tropp, J.A.: Greed is good: Algorithmic results for sparse approximation. IEEE Trans. Inf. Theory **50**(10), 2231–2242 (2004)
68. Tropp, J.A., Gilbert, A.C.: Signal recovery from random measurements via orthogonal matching pursuit. IEEE Trans. Inf. Theory **53**(12), 4655–4666 (2007)
69. Ugur, S., Arikan, O., Gürbüz, A.C.: Sar image reconstruction by expectation maximization based matching pursuit. Digit. Signal Process. **37**(1), 75–84 (2015)
70. Wallace, G.K.: The jpeg still picture compression standard. IEEE Trans. Consum. Electron. **38**(1), xviii–xxxiv (1992). https://doi.org/10.1109/30.125072
71. Wang, T.C., Liu, M.Y., Zhu, J.Y., Tao, A., Kautz, J., Catanzaro, B.: High-resolution image synthesis and semantic manipulation with conditional gans. In: arXiv preprint arXiv:1711.11585 (2017)
72. Wu, J., Liu, F., Jiao, L., Zhang, X., Hao, H., Wang, S.: Local maximal homogeneous region search for sar speckle reduction with sketch-based geometrical kernel function. IEEE Trans. Geosci. Remote Sens. **52**(9), 5751–5764 (2014). https://doi.org/10.1109/TGRS.2013.2292081
73. Wu, J., Liu, F., Jiao, L.C., Wang, X., Hou, B.: Multivariate compressive sensing for image reconstruction in the wavelet domain: using scale mixture models. IEEE Trans. Image Process. **20**(12), 3483–3494 (2011). https://doi.org/10.1109/TIP.2011.2150231
74. Xie, S., Tu, Z.: Holistically-nested edge detection. In: Proceedings of the IEEE International Conference on Computer Vision, pp. 1395–1403 (2015)
75. Xu, M., Li, C., Chen, Z., Wang, Z., Guan, Z.: Assessing visual quality of omnidirectional videos. IEEE Trans. Circuits Syst. Video Technol. **29**(12), 3516–3530 (2018). https://doi.org/10.1109/TCSVT.2018.2886277
76. Xu, M., Song, Y., Wang, J., Qiao, M., Huo, L., Wang, Z.: Predicting head movement in panoramic video: a deep reinforcement learning approach. IEEE Trans. Pattern Anal. Mach. Intell. **41**(11), 2693–2708 (2018). https://doi.org/10.1109/TPAMI.2018.2858783
77. Xu, M., Yang, L., Tao, X.M., Duan, Y.P., Wang, Z.L.: Saliency prediction on omnidirectional images with generative adversarial imitation learning (2019). https://arxiv.org/abs/1904.07080
78. Xu, W., Keshmiri, S., Wang, G.R.: Adversarially approximated autoencoder for image generation and manipulation. IEEE Trans. Multimedia **21**(9), 2387–2396. (2019). https://doi.org/10.1109/TMM.2019.2898777
79. Yang, S., Hu, Y., Yang, W., Duan, L.Y., Liu, J.: Towards coding for human and machine vision: scalable face image coding. IEEE Trans. Multimedia **23**, 2957–2971 (2021)

80. Yin, W., Osher, S., Goldfarb, D., Darbon, J.: Bregman iterative algorithms for l-minimization with applications to compressed sensing. SIAM J. Imag. Sci. **1**(1), 143–168 (2008)
81. Yu, G., Sapiro, G., Mallat, S.: Solving inverse problems with piecewise linear estimators: from gaussian mixture models to structured sparsity. IEEE Trans. Image Process. **21**(5), 2481–2499 (2012)
82. Zhang, Z., Xu, Y., Yang, J., Li, X., Zhang, D.: A survey of sparse representation: algorithms and applications. IEEE Access **3**, 490–530 (2017)
83. Zhao, J., Mathieu, M., Lecun, Y.: Energy-based generative adversarial network. In: International Conference on Learning Representations (2017)
84. Zhu, J., Park, T., Isola, P., Efros, A.A.: Unpaired image-to-image translation using cycle-consistent adversarial networks. In: 2017 IEEE International Conference on Computer Vision (ICCV), pp. 2242–2251 (2017). https://doi.org/10.1109/ICCV.2017.244

Chapter 5
End-to-End Semantic Information Transmission

5.1 Joint Semantic-Channel Coding for Multimodal Data

The proliferation of connected mobile devices and the increasing demand for intelligent capabilities have led to a significant surge in wireless data traffic. This trend poses new challenges for communication systems, such as facilitating diverse intelligent tasks, utilizing limited frequency resources, and handling massive volumes of data. Semantic communication, which selectively transmits relevant information, holds great promise in addressing these challenges by significantly reducing the required transmission resources [16, 18].

Traditional communication systems convert data into bits at the transmitter and rely on accurate bit recovery at the receiver, heavily dependent on good channel conditions and high signal-to-noise ratios (SNRs). Semantic communications transmit and recover the meaning of the conveyed content directly, eliminating the need for precise bit recovery and exhibiting robustness to channel impairments. Inspired by the emerging deep learning (DL) technologies, initial research on DL-enabled semantic communications focus on semantic recovery at the receiver for text [13, 22, 24], image [2], and speech [20]. However, there remains a significant gap in understanding how to effectively exploit semantic information for specific tasks. While some works have explored the use of semantic information in single-modal data, such as image classification [10] and image retrieval [7], the practical communication scenarios often involve the aggregation, transmission, and fusion of multimodal data from multiple users. This motivates our efforts in developing multimodal multiuser semantic communication systems.

Multimodal data refer to the representation of scenarios using different modalities. Typical examples of multimodal data include audio, video, images captured by electro-optical sensors, and text obtained from radio frequency sensors. In tasks that require the integration of multiple modalities, the multimodal data exhibit correlations within the contextual framework. Compared to situations involving single-modal data, multimodal data offer richer information for intelligent tasks,

X. Tao et al., *Wireless Multimedia Computational Communications*,
Wireless Networks, https://doi.org/10.1007/978-3-031-64155-8_5

introduce additional degrees of freedom, and improve the performance of intelligent tasks [5, 9]. Recent successful approaches for multimodal data fusion are mostly based on neural networks, and representative techniques include deep belief net (DBN), stacked autoencoder (SAE), convolutional neural network (CNN), and recurrent neural network (RNN) [5]. Multimodal semantic communications leverage multiple users to serve a single multimodal intelligent task, which is particularly relevant in emerging autonomous scenarios, i.e., autonomous checkout at retail stores. Establishing a multiuser semantic communication system to support multimodal data presents two main challenges:

- How can we extract the relevant semantic information from each user?
- How can we construct a model for multimodal semantic information fusion at the receiver?

5.1.1 System Model

As shown in Fig. 5.1, we consider the multiuser semantic communication system comprising a receiver with N_r antennas and N_U single-antenna transmitters. Our focus is multimodal multiuser communication. The multimodal multiuser scenario involves different users transmitting different types of data that are semantically complementary. Semantic complementarity refers to the situation where different multimodal data, such as images and text, provide mutually supportive information. For example, images are often accompanied by tags and textual explanations.

5.1.1.1 Semantic Transmitter

As shown in Fig. 5.1, we denote the source data of the ith user as s_i^Q, where $Q \subseteq \{\mathcal{I}: \text{image}, \mathcal{T}: \text{text}, \mathcal{V}: \text{video}, \mathcal{S}: \text{speech}\}$, where each source contains the semantic information. The semantic information is extracted first by:

$$z_i^Q = S\left(s_i^Q; \boldsymbol{\alpha}_i^Q\right), \tag{5.1}$$

where $z_i^Q \in \mathbb{R}^{L_S \times 1}$ represents the semantic information with length L_S[1] and $S\left(\cdot; \boldsymbol{\alpha}_i^Q\right)$ denotes the modality Q semantic encoder for the ith user with learnable parameters $\boldsymbol{\alpha}_i^Q$. Due to the limited resource and complex wireless communication environments, we compress the semantic information of the ith user using:

[1] The transmitted lengths of z_i^Q for different users could be different. To simplify the analysis, we choose the same length here for all users.

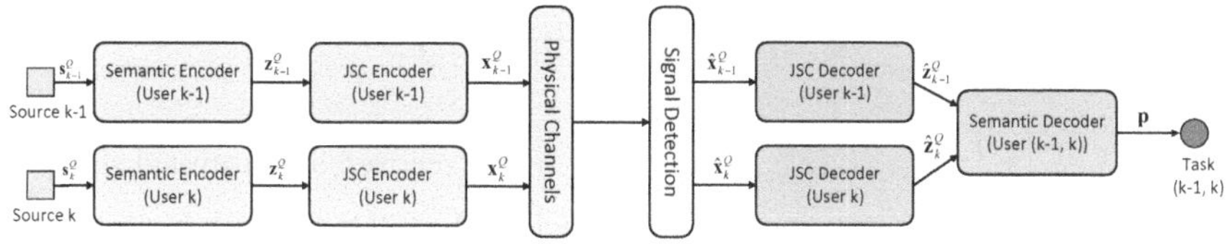

Fig. 5.1 The proposed framework for multiuser semantic communication systems

$$x_i^Q = C\left(z_i^Q; \boldsymbol{\beta}_i^Q\right), \tag{5.2}$$

where $x_i^Q \in \mathbb{C}^{L_C \times 1}$ denotes the transmitted complex signal with length $L_C < L_S$ and $C\left(\cdot; \boldsymbol{\beta}_i\right)$ represents the joint source-channel (JSC) encoder for modality Q by the ith user, with learnable parameters $\boldsymbol{\beta}_i$. The neural JSC encoder in semantic communication compresses the semantic information to reduce the number of transmitted symbols and enhance robustness to channel variations.

After the joint source-channel encoder, the signal power is normalized as:

$$\frac{1}{L_C}\mathbb{E}\left[\|\mathbf{x}_i^Q\|_2^2\right] \leq P, \tag{5.3}$$

where P is the power constraint for each transmitter.

5.1.1.2 Semantic Receiver

When the transmitted signal passes a multiple-input multiple-output (MIMO) physical channel, the received signal at the receiver, denoted as $\mathbf{Y} \in \mathbb{C}^{N_r \times L_C}$, can be expressed as:

$$\mathbf{Y} = \mathbf{H}\mathbf{X} + \mathbf{N}, \tag{5.4}$$

where $\mathbf{X}^T = \left[x_1^Q, x_2^Q, \ldots, x_{N_U}^Q\right] \in \mathbb{C}^{L_C \times K}$ denotes the transmit symbols from all N_U users, $\mathbf{H} = \left[\boldsymbol{h}_1, \boldsymbol{h}_2, \ldots, \boldsymbol{h}_{N_U}\right] \in \mathbb{C}^{N_r \times N_U}$ is the channel matrix between the base station (BS) and users. For the Rayleigh fading channel, the channel coefficients follow $\boldsymbol{h}_i \sim \mathcal{CN}(0, \mathbf{I}_{N_r})$; for the Rician fading channel, it follows $\boldsymbol{h}_i \sim \mathcal{CN}(\mu \mathbf{1}_{N_r \times 1}, \sigma^2 \mathbf{I}_{N_r})$ with $\mu = \sqrt{r/(r+1)}$ and $\sigma = \sqrt{1/(r+1)}$, where $\mathbf{I}_{N_r}$ represents the $N_r \times N_r$ identity matrix, $\mathbf{1}_{N_r \times 1}$ denotes the all-one vector with length N_r, and r is the Rician coefficient. $\mathbf{N} \in \mathbb{C}^{N_r \times L_C}$ represents the circular symmetric Gaussian noise. The elements of $\mathbf{N}$ are independent and identically distributed (i.i.d) with zero mean and variance σ^2, and SNR is $\sum_i \left\|\boldsymbol{h}_i x_i^Q\right\|^2 / \sigma^2$.

Subsequently, the transmitted signals are recovered by the linear minimum mean squared error (L-MMSE) detector with estimated channel state information (CSI):

$$\widehat{\mathbf{X}} = \widehat{\mathbf{H}}^{\mathrm{H}}\left(\widehat{\mathbf{H}}\widehat{\mathbf{H}}^{\mathrm{H}} + \sigma^2\mathbf{I}_{\mathrm{sem}}\right)^{-1}\mathbf{Y}, \tag{5.5}$$

where $\widehat{\mathbf{X}}^T = \left[\hat{\boldsymbol{x}}_1^Q; \hat{\boldsymbol{x}}_2^Q; \cdots; \hat{\boldsymbol{x}}_{N_U}^Q\right] \in \mathbb{C}^{L_C \times N_U}$ is the recovered transmitted signals, $\widehat{\mathbf{H}} = \mathbf{H} + \Delta\mathbf{H}$ is the estimated CSI, in which $\Delta\mathbf{H} \in \mathcal{CN}(0, \sigma_e^2)$ represents the estimation error, and σ_e^2 measures the accuracy of the channel estimation.

The semantic information from the ith user, $\hat{z}_i^Q \in \mathbb{R}^{L_S \times 1}$, is recovered by the JSC decoder as:

$$\hat{z}_i^Q = C^{-1}\left(\hat{\boldsymbol{x}}_i^Q; \boldsymbol{\gamma}_i^Q\right), \tag{5.6}$$

where $C^{-1}\left(\hat{\boldsymbol{x}}_i^Q; \boldsymbol{\gamma}_i^Q\right)$[2] represents JSC decoder for the ith user and the modality Q, with the learned parameters $\boldsymbol{\gamma}_i^Q$. The JSC decoder aims to decompress the semantic information while mitigating the effects of channel distortion and inter-user interference. For multimodal tasks, a multimodal semantic receiver is employed.

To perform the final task using the multimodal semantic information, we merge the semantic information from different users:

$$\boldsymbol{p} = S^{-1}\left(\hat{z}_1^Q, \hat{z}_2^Q, \ldots, \hat{z}_{N_U}^Q; \boldsymbol{\varphi}^{\mathbf{sem}}{}_{(1,2,\ldots,N_U)}\right), \tag{5.7}$$

where $\boldsymbol{p}$ represents the results of the multimodal task and $S^{-1}\left(\cdot; \boldsymbol{\varphi}^{\mathbf{sem}}{}_{(1,2,\ldots,N_U)}\right)$ denotes the multimodal semantic decoder with learnable parameters $\boldsymbol{\varphi}^{\mathbf{sem}}{}_{(1,2,\ldots,N_U)}$.

5.1.2 *Multimodal Multiuser Semantic Communications*

In this section, we investigate multimodal multiuser semantic communications for the visual question answering (VQA) task, specifically a novel approach called DeepSC-VQA. We propose a semantic decoder to merge image-text semantic information.

5.1.2.1 Model Description

Let us assume the ith user is responsible for image transmission, and the uth user is responsible for text transmission. The training dataset is denoted as $\mathcal{D}_{i,j}^{\mathcal{I},\mathcal{T}} = \left\{(\boldsymbol{s}_{i,k_{it}}^{\mathcal{I}}, \boldsymbol{s}_{j,k_{it}}^{\mathcal{T}}, l_{(i,j),k_{it}})\right\}_{k_{it}=1}^{D}$ with size D. Here, $\boldsymbol{s}_{i,k_{it}}^{\mathcal{I}}$ represents the k_{im}-th image

[2] In order to reduce the number of representation symbols, we use $\cdot^{-1}$ here to represent the decoder.

from the ith user, $s^{\mathcal{T}}_{j,k_{it}}$ represents the k_{it}-th text from the uth user, and $l_{(i,j),k_{it}}$ is the answer label for $s^{\mathcal{I}}_{i,k_{it}}$ and $s^{\mathcal{T}}_{j,k_{it}}$. $S_{\text{VQA}}\left(\cdot;\boldsymbol{\alpha}_i^{\mathcal{I}}\right)$, $C_{\text{VQA}}\left(\cdot;\boldsymbol{\beta}_i^{\mathcal{I}}\right)$, and $C_{\text{VQA}}^{-1}\left(\cdot;\boldsymbol{\gamma}_i^{\mathcal{I}}\right)$ represent the image semantic encoder, image JSC encoder, and image JSC decoder of the ith user, respectively. Similarly, $S_{\text{VQA}}\left(\cdot;\boldsymbol{\alpha}_j^{\mathcal{T}}\right)$, $C_{\text{VQA}}\left(\cdot;\boldsymbol{\beta}_j^{\mathcal{T}}\right)$, and $C_{\text{VQA}}^{-1}\left(\cdot;\boldsymbol{\gamma}_j^{\mathcal{T}}\right)$ represent the text semantic encoder, text JSC encoder, and text JSC decoder of the uth user, respectively. $S_{\text{VQA}}^{-1}\left(\cdot;\boldsymbol{\varphi}^{\textbf{sem}}{}_{(i,j)}\right)$ represents joint semantic decoder of the ith and the uth user for the VQA task.

As shown in Fig. 5.2, the proposed DeepSC-VQA network consists of an image transmitter, a text transmitter, and a receiver for simplicity. For the DeepSC-VQA transmitters and receivers, the image transmitter comprises an image semantic encoder for extracting image semantic information to be transmitted and a JSC encoder for compressing the semantic information. The semantic encoder includes multiple vision transformer layers, and the JSC encoder employs fully connected layers with varying units. The vision transformer is superior to CNNs in capturing global features and exhibits increased robustness to input image distortions. After transmitting and performing signal detection, the receiver employs the JSC decoder with different units to recover the transmitted image semantic information. The text transmitter includes a text semantic encoder and a text JSC encoder for extracting and compressing the semantic text information, respectively. The text

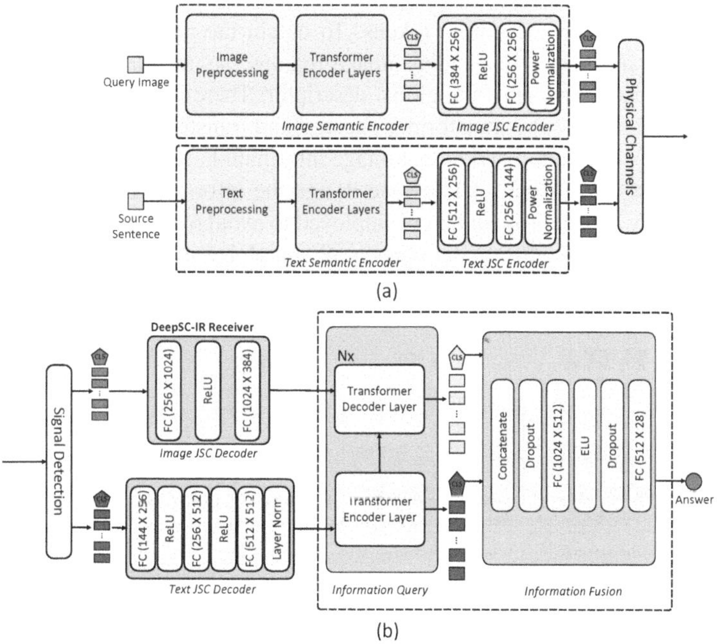

Fig. 5.2 The proposed network structure of multimodal multiuser semantic communication system with DeepSC-VQA transceiver. (**a**) Transmitter. (**b**) Receiver

semantic encoder adopts multiple transformer encoder layers, and the text JSC encoder is specially designed. Compared to RNNs, the transformer is better at capturing relationships between sentences and can be trained significantly faster than architectures based on recurrent layers. The designed text JSC decoder at the receiver recovers the semantic text information from distorted signals. Furthermore, we introduce a new semantic decoder network for image-text information fusion, which comprises two modules: the information query module and the information fusion module.

Information Query

Images and text possess different modalities, and each modality can provide semantic complementary information to the other. To exploit this semantic complementarity, we adopt a layer-wise transformer. Figure 5.3 depicts the comparison between the classic transformer and the layer-wise transformer. Unlike the classic transformer, where the decoder layers use the output tokens of the last layer of the encoder as the input, the layer-wise transformer employs the output tokens of each encoder layer as input for each decoder layer. This design generates more text information than the classic transformer and guides the more efficient image information query in the decoder, without incurring any additional costs.

Information Fusion

After the information query, the layer-wise transformer has captured keywords in the text information and the corresponding regions in the image information, which are reflected in the output tokens. To obtain the answer, we need to fuse the keywords and the corresponding image regions. As mentioned in Section II, the <CLS> token represents the global descriptor. Therefore, the <CLS> tokens in the output tokens of the transformer encoder and transformer decoder represent the global text information and global image information, respectively. By utilizing the text<CLS> and image <CLS>, we design the information fusion module as shown in Fig. 5.2, with dropout layers employed to avoid overfitting. In comparison to employing descriptor fusion networks to obtain global descriptors, our proposed

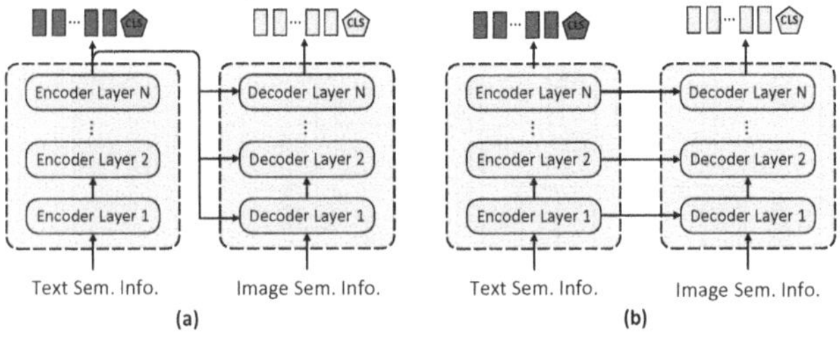

Fig. 5.3 Comparison between (**a**) classic transformer and (**b**) layer-wise transformer

information fusion approach directly uses the <CLS> global descriptors and achieves similar answer accuracy without requiring additional descriptor fusion networks.

5.1.2.2 Training Algorithm

As shown in Algorithm 1, DeepSC-VQA is trained through three phases with different loss functions.

The first phase is `Train Semantic Codec`, where the semantic codec of DeepSC-VQA, $S_{\text{VQA}}\left(\cdot;\boldsymbol{\alpha}_i^{\mathcal{I}}\right)$, $S_{\text{VQA}}\left(\cdot;\boldsymbol{\alpha}_j^{\mathcal{T}}\right)$, $S_{\text{VQA}}^{-1}\left(\cdot;\boldsymbol{\varphi}^{\text{sem}}{}_{(i,j)}\right)$, is jointly trained using the CE loss function:

$$\mathcal{L}_{\text{VQA}} = \mathbb{E}\left[-P\left(l_{(i,j),k_{it}}\right)\log\left(P\left(\hat{l}_{(i,j),k_{it}}\right)\right)\right], \tag{5.8}$$

where $P(l_{(i,j),k_{it}})$ and $P(\hat{l}_{(i,j),k_{it}})$ represent the real and predicted probabilities of the answer, respectively. By minimizing the CE loss, the network learns to predict the answer with the highest probability of accuracy.

After training the semantic codec, the JSC codecs are trained to compress the semantic information using the JSC encoder and then decompress it accurately over multiple user physical channels using the JSC decoder. The image and text JSC codecs are jointly trained through the `Train JSC Codec` function, with the loss function defined as:

$$\mathcal{L}_{\text{MSE}}^{(\text{VQA})} = \mathbb{E}\left[\left\|\hat{z}_{i,k_{it}}^{\mathcal{I}} - z_{i,k_{it}}^{\mathcal{I}}\right\|_2^2 + \left\|\hat{z}_{j,k_{it}}^{\mathcal{T}} - z_{j,k_{it}}^{\mathcal{T}}\right\|_2^2\right], \tag{5.9}$$

where $z_{i,k_{it}}^{\mathcal{I}}$ and $z_{j,k_{it}}^{\mathcal{T}}$ represent the transmitted semantic image and text information, respectively. $\hat{z}_{i,k_{it}}^{\mathcal{I}}$ and $\hat{z}_{j,k_{it}}^{\mathcal{T}}$ represent the recovered semantic image and text information at the receiver, respectively.

Due to error propagation from the JSC decoders to the semantic receiver, resulting from imperfect semantic information recovery in low SNR regimes, the entire DeepSC-VQA network is jointly trained with the loss function (5.8) to minimize error propagation. This is accomplished through the `Train Whole Network` function.

5.1.3 Simulation Results

In this section, we compare the performance of the proposed multiuser semantic communication systems with traditional source coding and channel coding methods over various channels, in which both the perfect and imperfect CSI are considered.

Algorithm 1: DeepSC-VQA training algorithm

Data: The training dataset $\mathcal{D}_i$.

1 **Function** `Train Semantic Coder()`:

  Input: Choose mini-batch data $\left\{(\boldsymbol{s}_{i,k_{it}}, \boldsymbol{s}_{j,k_{it}}, l_{(i,j),k_{it}})\right\}_{k_{it}=n}^{n+B}$ from $\mathcal{D}_{(i,j)}^{\mathrm{I,\,T}}$.

2 **for** $k_{it} = n \to n + B$ **do**

3   $S_{\mathtt{VQA}}\left(\boldsymbol{s}_{i,k_{it}}^{\mathcal{I}}; \boldsymbol{\alpha}_i^{\mathcal{I}}\right) \to \boldsymbol{z}_{i,k_{it}}^{\mathcal{I}}$,

4   $S_{\mathtt{VQA}}\left(\boldsymbol{s}_{j,k_{it}}^{\mathcal{T}}; \boldsymbol{\alpha}_j^{\mathcal{T}}\right) \to \boldsymbol{z}_{j,k_{it}}^{\mathcal{T}}$,

5   $S_{\mathtt{VQA}}^{-1}\left(\boldsymbol{z}_{i,k_{it}}^{\mathcal{I}}, \boldsymbol{z}_{j,k_{it}}^{\mathcal{T}}; \boldsymbol{\varphi^{\mathrm{sem}}}_{(i,j)}\right) \to \hat{l}_{(i,j),k_{it}}$,

6   Compute $\mathcal{L}_{\mathtt{VQA}}$ by (5.8) with $l_{(i,j),k_{it}}$ and $\hat{l}_{(i,j),k_{it}}$.

7 **end**

8 Train $\boldsymbol{\alpha}_i^{\mathcal{I}}, \boldsymbol{\alpha}_j^{\mathcal{T}}, \boldsymbol{\varphi^{\mathrm{sem}}}_{(i,j)} \to$ Gradient descent with $\mathcal{L}_{\mathtt{VQA}}$

  Return: $S_{\mathtt{VQA}}\left(; \boldsymbol{\alpha}_i^{\mathcal{I}}\right)$, $S_{\mathtt{VQA}}\left(; \boldsymbol{\alpha}_j^{\mathcal{T}}\right)$, and $S_{\mathtt{VQA}}^{-1}\left(; \boldsymbol{\varphi^{\mathrm{sem}}}_{(i,j)}\right)$.

9 **Function** `Train Channel Coder()`:

  Input: The semantic text and image information, $\left\{\boldsymbol{z}_{i,k_{it}}^{\mathcal{I}}, \boldsymbol{z}_{j,k_{it}}^{\mathcal{T}}\right\}_{k_{it}=n}^{n+B}$.

10 **for** $k_{it} = n \to n + B$ **do**

11   **Transmitter**:

12    $C_{\mathtt{VQA}}\left(\boldsymbol{z}_{i,k_{it}}^{\mathcal{I}}; \boldsymbol{\beta}_i^{\mathcal{I}}\right) \to \boldsymbol{x}_{i,k_{it}}^{\mathcal{I}}$, $C_{\mathtt{VQA}}\left(\boldsymbol{z}_{j,k_{it}}^{\mathcal{T}}; \boldsymbol{\beta}_j^{\mathcal{T}}\right) \to \boldsymbol{x}_{j,k_{it}}^{\mathcal{T}}$,

13   Transmit $\boldsymbol{x}_{i,k_{it}}^{\mathcal{I}}, \boldsymbol{x}_{j,k_{it}}^{\mathcal{T}}$ over the air.

14   **Receiver**:

15   Receive $\mathbf{Y}$,

16   MIMO detection by (5.5) to get $\hat{\boldsymbol{x}}_{i,k_{it}}^{\mathcal{I}}$ and $\hat{\boldsymbol{x}}_{j,k_{it}}^{\mathcal{T}}$,

17    $C_{\mathtt{VQA}}^{-1}\left(\hat{\boldsymbol{x}}_{i,k_{it}}^{\mathcal{I}}; \boldsymbol{\gamma}_i^{\mathcal{I}}\right) \to \hat{\boldsymbol{z}}_{i,k_{it}}^{\mathcal{I}}$, $C_{\mathtt{VQA}}^{-1}\left(\hat{\boldsymbol{x}}_{j,k_{it}}^{\mathcal{T}}; \boldsymbol{\gamma}_j^{\mathcal{T}}\right) \to \hat{\boldsymbol{z}}_{j,k_{it}}^{\mathcal{T}}$,

18   Compute $\mathcal{L}_{\mathtt{MSE}}$ by (5.9) with $\boldsymbol{z}_{i,k_{it}}^{\mathcal{I}}, \boldsymbol{z}_{j,k_{it}}^{\mathcal{T}}, \hat{\boldsymbol{z}}_{i,k_{it}}^{\mathcal{I}}$, and $\hat{\boldsymbol{z}}_{j,k_{it}}^{\mathcal{T}}$.

19 **end**

20 Train $\boldsymbol{\beta}_i^{\mathcal{I}}, \boldsymbol{\gamma}_i^{\mathcal{I}}, \boldsymbol{\beta}_j^{\mathcal{T}}, \boldsymbol{\gamma}_j^{\mathcal{T}} \to$ Gradient descent with $\mathcal{L}_{\mathtt{MSE}}$,

  Return: $C_{\mathtt{VQA}}\left(; \boldsymbol{\beta}_i^{\mathcal{I}}\right)$, $C_{\mathtt{VQA}}\left(; \boldsymbol{\beta}_j^{\mathcal{T}}\right)$, $C_{\mathtt{VQA}}^{-1}\left(; \boldsymbol{\gamma}_i^{\mathcal{I}}\right)$, and $C_{\mathtt{VQA}}^{-1}\left(; \boldsymbol{\gamma}_j^{\mathcal{T}}\right)$.

21 **Function** `Train Whole Network()`:

  Input: Choose mini-batch data $\left\{(\boldsymbol{s}_{i,k_{it}}^{\mathcal{I}}, \boldsymbol{s}_{j,k_{it}}^{\mathcal{T}}, l_{(i,j),k_{it}})\right\}_{k_{it}=n}^{n+B}$ from $\mathcal{D}_{(i,j)}^{\mathcal{I},\mathcal{T}}$.

22 **for** $k_{it} = n \to n + B$ **do**

23   Repeat line 3–4, 12–19, and 5 to get $\hat{l}_{(i,j),k_{it}}$,

24   Compute $\mathcal{L}_{\mathtt{VQA}}$ by (5.8) with $l_{(i,j),k_{it}}$ and $\hat{l}_{(i,j),k_{it}}$.

25 **end**

26 Train DeepSC-VQA $\to$ Gradient descent with $\mathcal{L}_{\mathtt{VQA}}$.

  Return: $S_{\mathtt{VQA}}\left(; \boldsymbol{\alpha}_i^{\mathcal{I}}\right)$, $S_{\mathtt{VQA}}\left(; \boldsymbol{\alpha}_j^{\mathcal{T}}\right)$, $S_{\mathtt{VQA}}^{-1}\left(; \boldsymbol{\varphi^{\mathrm{sem}}}_{(i,j)}\right)$,

   $C_{\mathtt{VQA}}\left(; \boldsymbol{\beta}_i^{\mathcal{I}}\right)$, $C_{\mathtt{VQA}}\left(; \boldsymbol{\beta}_j^{\mathcal{T}}\right)$, $C_{\mathtt{VQA}}^{-1}\left(; \boldsymbol{\gamma}_i^{\mathcal{I}}\right)$, and $C_{\mathtt{VQA}}^{-1}\left(; \boldsymbol{\gamma}_j^{\mathcal{T}}\right)$.

5.1.3.1 Implementation Details

The Datasets

For the VQA task, we adopt the popular dataset *CLEVR*, which consists of a training set of 70,000 images and 699,989 questions, as well as a test set of 15,000 images and 149,991 questions.

Training Settings

The image semantic encoder in DeepSC-VQA is based on the pretrained DeiT small model, while the remaining parts are trained from scratch. The text semantic encoder consists of six transformer encoder layers, and the semantic decoder comprises four transformer encoder layers and decoder layers. We freeze the image semantic encoder to expedite training. The output dimensions for the vision transformer and text transformer are set differently, which require dimension-increasing operations after the image JSC decoder. The dimension-increasing operations involve a dropout layer, a dense layer from 384 to 512, an exponential linear unit (ELU) activation layer, another dropout layer, and a dense layer from 512 to 512, followed by an ELU activation layer. The setting of the `Train Semantic Codec` in DeepSC-VQA employs the Adam optimizer with a learning rate 1×10^{-4}, betas of 0.9 and 0.98, batch size of 64, and 80 epochs. The setting of the `Train JSC Codec` in DeepSC-VQA uses the Adam optimizer with a learning rate of 1×10^{-3}, batch size of 128, and 30 epochs. The setting of `Train Whole Network` of DeepSC-VQA is the same as that of the `Train Semantic Codec` but with 10 epochs. Data augmentation is employed to resize images to 224×224 using BICUBIC interpolation for both training and testing.

5.1.3.2 Benchmarks and Performance Metrics

Our benchmarks include several typical source and channel coding methods:

- Error-free transmission: Noiseless images and texts are delivered to the receiver, representing the upper bound.
- Semi-conventional method: Semantic information, extracted by the semantic encoder, is transmitted using conventional separate source-channel coding, employing the following technologies:
  - 8-bit μ-law quantization for mapping semantic information into bits
  - Low-density parity-check code (LDPC) for channel coding
  - 8-quadrature amplitude modulation (QAM) for modulation
- Conventional methods: Source and channel coding are performed separately using the following technologies:
  - 8-bit unicode transformation format (UTF-8) encoding for text source coding, a commonly used method in text compression

  - Joint photographic experts group (JEPG) for image source coding, a widely used method in image compression
  - Turbo coding for text channel coding, popular channel coding for a small size file
  - LDPC for image channel coding and classic channel coding for big size files.
- Hybrid methods: Multimodal data are transmitted using a combination of conventional methods and the proposed DeepSC. We use the following technologies:
  - UTF-8 and turbo coding for text transmission and DeepSC-VQA for image transmission
  - JPEG and LDPC coding for image transmission and DeepSC-VQA for text transmission

In the simulation, the coding rate is 1/3 and the block length is 256 for the turbo codes. The LDPC codes employ DVB-S.2 standard, with a coding rate of 1/3, a parity-check matrix size of 43,200 $\times$ 64,800, and a block length of 64,800. We employ the LS channel estimation, in which perfect and imperfect CSI are considered with $\sigma_e^2 = 0$ and $\sigma_e^2 = \sigma^2$, respectively. The Rician channels employ $r = 2$, and the additive white Gaussian noise (AWGN) channels have $\mathbf{H} = \mathbf{I}$. The coherent time is set as the transmission time for each batch in the simulation. For metrics versus SNRs, we set $N_r = K = 2$, and for metrics versus different numbers of users, $N_r = K > 2$.

5.1.3.3 Multimodal Multiuser Semantic Communications

Figure 5.4 presents the comparison of answer accuracy performance for the VQA task over different channels. The benchmark includes the conventional method with UTF-8-Turbo using BPSK for text and JPEG-LDPC with 8-QAM for image, as well as two hybrid methods. DeepSC-VQA outperforms the benchmark in the low SNR regimes over AWGN channels and outperforms all benchmarks in all SNR regimes over fading channels. Notably, DeepSC-VQA achieves the upper bound at an approximate SNR of 9 dB over fading channels. While the benchmarks experience a significant decrease in answer accuracy from AWGN to fading channels, DeepSC-VQA only experiences minimal performance degradation in low SNR regimes and no performance loss in high SNR regimes. Moreover, replacing one of the conventional methods (i.e., UTF-8 or JPEG) with DeepSC-VQA achieves higher answer accuracy in all SNR regimes compared to JPEG-UTF-8. This suggests that employing semantic information can enhance the robustness of multimodal data transmission. For dynamic imperfect CSI in Fig. 5.4c, DeepSC-VQA exhibits superior robustness compared to all benchmarks and outperforms JPEG-UTF-8 with a gain of 16 dB at 0.6 answer accuracy. Similarly, transmitting semantic information improves answer accuracy for dynamic imperfect CSI compared to JPEG-UTF-8.

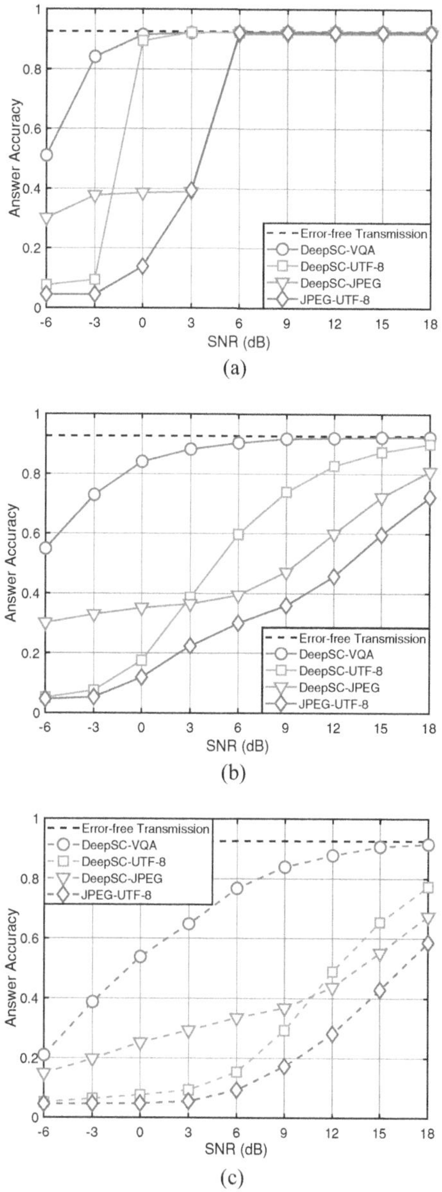

Fig. 5.4 Answer accuracy comparison between DeepSC-VQA, conventional methods, and hybrid methods, including UTF-8-Turbo with BPSK for text and JPEG-LDPC with 8-QAM for image, in which different channels are considered. (**a**) AWGN channels. (**b**) Rician channels with perfect CSI. (**c**) Rician channels with imperfect CSI

Table 5.1 demonstrates the comparison of answer accuracy between the classic transformer and the layer-wise transformer trained with 50 epochs. From Table 5.1, the layer-wise transformer with the proposed fusion method outperforms the classic transformer with the standard fusion method by 37.4% in terms of answer accuracy. This further confirms the effectiveness of DeepSC-VQA.

Table 5.1 Answer accuracy comparison between the layer-wise transformer–based multimodal fusion method and the classic transformer–based standard fusion method

Classic transformer with classic fusion	Classic transformer with proposed fusion
55.1%	57.3%
Layer-wise transformer with classic fusion	Layer-wise transformer with proposed fusion
92.5%	92.3%

Table 5.2 The number of transmitted symbols comparison between multiuser semantic communication systems and traditional source-channel communication systems

Modality	Methods	# of transmitted symbols	Ratio
Text	DeepSC-VQA / UTF-8-Turbo with BPSK	77/152	50.66%
Image	DeepSC-VQA / JPEG-LDPC with 8-QAM	25, 216/55, 624	45.33%

Table 5.3 Computational complexity comparison between multiuser semantic communication systems and traditional source-channel communication systems

		Computational complexity	
Dataset	Methods	Additions	Multiplications
Text	DeepSC-VQA / UTF-8-Turbo with BPSK	$5.9 \times 10^5/1.0 \times 10^5$	$5.9 \times 10^5/1.6 \times 10^5$
Image	DeepSC-VQA / JPEG-LDPC with 8-QAM	$1.6 \times 10^8/1.0 \times 10^9$	$1.6 \times 10^8/1.9 \times 10^9$

5.1.3.4 Latency and Computational Complexity Analysis

Table 5.2 compares the number of transmission symbols for different methods. The proposed multiuser semantic communication systems significantly reduce the number of transmission symbols for image transmission. For text transmission, they achieve approximately 50% reduction in the number of symbols when the benchmark employs a lower-order modulation in the VQA task. This indicates that the proposed multiuser semantic communications can decrease transmission delay with a lower number of symbols, making them suitable for low-latency scenarios.

Table 5.3 compares the computational complexity of different methods. We focus on analyzing the complexity of channel coding because the other parts are shared between both methods, and the complexity of source coding is low and can be neglected. The computational complexity of the proposed method primarily depends on matrix multiplication. For the traditional method, we mainly calculate the decode complexity per bit, following the computation method shown in [25]. Regarding image transmission, the proposed method exhibits lower computational complexity than traditional methods. For text transmission, the proposed method demonstrates similar computational complexity. This indicates that the proposed multiuser semantic communication systems achieve lower power consumption when transmitting large-sized data.

5.1.4 *Conclusions*

In this section, we have explored task-oriented multiuser semantic communications for transmitting data with multiple modalities. We considered one multimodal task, VQA. In this context, we have proposed DeepSC-VQA, a transformer-based transceiver, trained jointly using the proposed training algorithm. In addition, the proposed multiuser semantic communication systems outperform traditional systems in low SNR regimes and exhibit graceful performance degradation with imperfect CSI. DeepSC-VQA provides more than 18 dB gain, reduces the number of transmission symbols by over 50%, and decreases computational complexity compared to traditional communications. Based on these findings, we conclude that multiuser semantic communication systems are an attractive alternative to traditional systems for specific tasks.

5.2 Semantic-Aware Network Resource Allocation

In Sect. 5.1, we have observed that semantic communications, which transmit task-related information, have the potential to improve transmission reliability and efficiency, particularly in the low signal-to-noise ratio (SNR) regime. Different types of tasks, including single-modal tasks (e.g., text transmission [22] and image retrieval [7]) and multimodal tasks (e.g., VQA [23]), require different semantic transceivers to achieve superior performance. Therefore, the conventional resource allocation model is not applicable, and it is crucial to investigate resource allocation for such semantic-aware networks to enhance communication efficiency while ensuring transmission reliability.

Several preliminary studies have been conducted on semantic-aware resource allocation. Researchers in [19] and [11] considered the importance of weights of the extracted semantic features to formulate the problem. However, both works aimed to maximize the task performance, without evaluating the semantic communication efficiency. Although the authors in [21] defined the system throughput in terms of message for text transmission and proposed a method to maximize this performance, they optimized the resource allocation at the message level rather than the semantic level.

To address the following challenges in semantic-level resource allocation:

- How to evaluate the transmission efficiency at the semantic level?
- How to model multimodal tasks and handle the coexistence of multiple tasks?
- How to formulate the resource allocation at the semantic level?

We propose to investigate resource allocation for a single-task network to tackle the first and third challenges. Subsequently, we propose semantic-aware resource allocation for a multitask network to address the second and third challenges.

5.2.1 Resource Allocation for a Single-Task Network

5.2.1.1 System Model

We consider a cellular network consisting of a BS and a set of users denoted as $\mathcal{U} = \{1, 2, \ldots, i, \ldots, N_U\}$, as depicted in Fig. 5.5. DeepSC [22], a semantic communication model based on transformer is employed at each user for text transmission. Transformer enables effective extraction of semantic information underlying the text. The DeepSC transceiver is assumed to be trained at the BS or cloud platforms, and the trained semantic transmitter model is broadcasted to users. We will elaborate on the DeepSC transmitter at users, the transmission model, and the DeepSC receiver at the BS.

(1) DeepSC Transmitter

In the considered model, the ith user generates a sentence $\mathbf{s}_i = [w_{i,1}, w_{i,2}, \ldots, w_{i,o}, \ldots, w_{i,O_i}]$, where $w_{i,o}$ represents the oth word and O_i is the sentence length at the ith user. The sentence is then input to the DeepSC transmitter, which maps it to a semantic symbol vector $\mathbf{X}_i = [\mathbf{x}_{i,1}, \mathbf{x}_{i,2}, \ldots, \mathbf{x}_{i,n_i^w O_i}]$, where $\mathbf{X}_i \in \mathbb{R}^{n_i^w O_i \times 2}$ and $n_i^w O_i$ denote the length of the semantic symbol vector for a sentence at the ith user. Notably, the length of $\mathbf{X}_i$ varies with O_i to effectively extract the semantic information from sentences of different lengths more effectively [22]. In this model, n_i^w represents the average number of semantic symbols used for each word at the ith user, and each semantic symbol can be directly transmitted over the transmission medium.

(2) Transmission Model

Let $\mathcal{K} = \{1, 2, \ldots, k, \ldots, N_K\}$ denote the set of available channels in the network, where N_k is the number of channels and each channel has a bandwidth B_W. The channel assignment vector of the ith user is denoted as $\boldsymbol{\alpha}_i =$

Fig. 5.5 The structure of the single-task semantic-aware network

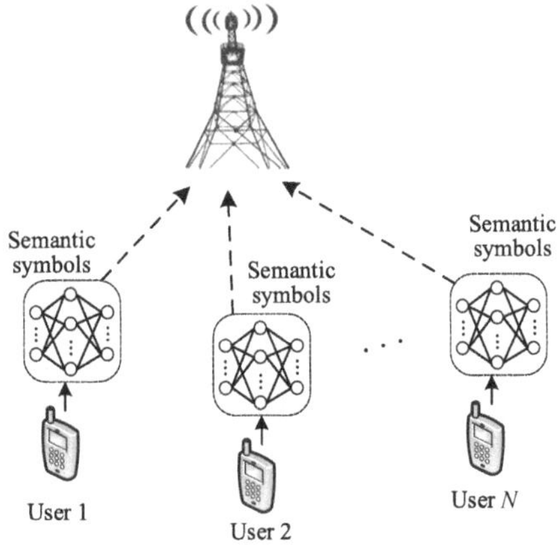

$\{\alpha_{i,1}, \alpha_{i,2}, \ldots, \alpha_{i,k}, \ldots, \alpha_{i,N_K}\}$, where $\alpha_{i,k} \in \{0, 1\}$. Specifically, $\alpha_{i,k} = 1$ indicates that the kth channel is allocated to the ith user, while $\alpha_{i,k} = 0$ indicates otherwise. We assume that each channel can only be allocated to at most one user, and each user can only occupy at most one channel. Hence, the following conditions hold:

$$\sum_{i=1}^{N_U} \alpha_{i,k} \leq 1, \ \forall k \in \mathcal{K}, \tag{5.10}$$

and

$$\sum_{k=1}^{N_K} \alpha_{i,k} \leq 1, \ \forall i \in \mathcal{U}. \tag{5.11}$$

In addition, we consider that all channels consist of large-scale fading and small-scale Rayleigh fading. The SNR of the ith user over the kth channel is given by:

$$\gamma_{i,k} = \frac{p_i g_i |h_{i,k}|^2}{B_W \sigma^2}, \tag{5.12}$$

where p_i is the transmit power of the ith user, g_i is the large-scale channel gain of the ith user (including path loss and shadowing), $h_{i,k} \sim \mathcal{CN}(0, 1)$ is the Rayleigh fading coefficient for the ith user transmitting over the kth channel, and σ^2 is the noise power spectral density.

(3) DeepSC Receiver

At the BS, the signal received from the ith user can be expressed as $\mathbf{Y}_i = \sqrt{g_i} h_{i,k} \mathbf{X}_i + \mathbf{z}$, where $\mathbf{z}$ is AWGN, and each element of $\mathbf{z}$ follows $\mathcal{CN}(0, \sigma^2)$. The received signal is first decoded by the channel decoder and then the semantic decoder to estimate sentence $\hat{\mathbf{s}}_i$. To evaluate the performance of semantic communications for text transmission, we adopt the semantic similarity [22] as the performance metric:

$$\xi = \frac{\mathbf{B}(s)\mathbf{B}(\hat{s})^{\mathrm{T}}}{\|\mathbf{B}(s)\| \, \|\mathbf{B}(\hat{s})\|}, \tag{5.13}$$

where $\mathbf{B}(\cdot)$ denotes sentence-bidirectional encoder representations from the transformers (BERT) model. This model exhibits significant improvement over state-of-the-art sentence embedding methods. A pretrained sentence-BERT model [17] is adopted. Compared to other semantic metrics like bilingual evaluation understudy (BLEU) [14], BERT-level similarity provides a more precise measure of the semantic distance between two sentences. From Eq. (5.13), we have $0 \leq \xi \leq 1$, where $\xi = 1$ indicates that two sentences have the highest similarity, while $\xi = 0$ signifies no similarity between them.

5.2.1.2 Resource Allocation Scheme

In this part, the semantic spectral efficiency (S-SE) is first defined as a new metric for semantic-aware networks. Then the semantic-aware resource allocation is formulated as an S-SE maximization problem considering channel assignment and the number of transmitted semantic symbols. Finally, the optimal solution to the optimization problem is obtained.

(1) Semantic Spectral Efficiency

In conventional communications, spectral efficiency is measured in bits per second per Hertz (*bits/s/Hz*), which is suitable for measuring the transmission rate of bit sequences but not semantic information. This is because bit sequences are produced based on statistical knowledge of the source and are unrelated to the meaning of the source. Therefore, new performance metrics need to be explored at the semantic level.

For clarity, we assume that semantic information can be measured by the *semantic unit* (*sut*), which represents the basic unit of semantic information.[3] Based on this, we can define two crucial semantic-based performance metrics:

- *Semantic transmission rate (S-R)* refers to the effectively transmitted semantic information per second and is measured in *suts/s*.
- *S-SE* refers to the rate at which semantic information can be successfully transmitted over a unit of bandwidth and is measured in *suts/s/Hz*.

Next, we derive the expressions for S-R and S-SE. Let $\mathcal{D}^{\text{text}} = \{(\mathbf{s}_u = [w_{u,1}, w_{u,2}, \ldots, w_{u,o_u}, \ldots, w_{u,O_u}])\}_{u=1}^{D^{\text{text}}}$ with size D^{text} representing the text dataset, where $\mathbf{s}_u$ is the uth sentence with a length of O_u and $w_{u,o}$ represents the oth word. The amount of semantic information in $\mathbf{s}_u$ is denoted as I_u. With $p(\mathbf{s}_u)$ representing the occurrence probability of $\mathbf{s}_u$, the expected amount of semantic information per sentence can be expressed as $I_{\text{sem}} = \sum_{u=1}^{D^{\text{text}}} I_u p(\mathbf{s}_u)$, which corresponds to an expected number of words per sentence as $O = \sum_{u=1}^{D^{\text{text}}} O_u p(\mathbf{s}_u)$. Since we focus on the long-term text transmission rather than the transmission of individual sentences, the expected values I_{sem} and O, rather than random values, should be used to represent S-R and S-SE. Thus, at the ith user, on average, there are $n_i^w O_i$ semantic symbols on average carrying the amount of semantic information of I_{sem}, and the average amount of semantic information per semantic symbol is $I_{\text{sem}}/(n_i^w O_i)$. Moreover, since the symbol rate is equal to the channel bandwidth for passband transmission, the total semantic information transmitted over a channel with bandwidth B_W is $B_W I_{\text{sem}}/(n_i^w O_i)$. Thus, the S-R of the ith user over the kth channel can be expressed as:

[3] The *semantic unit* here is just a concept and will not affect the resource optimization solution, the reason of which will be clarified later in this section.

$$\Gamma_{i,k} = \frac{B_W I_{\mathrm{sem}}}{n_i^w O_i} \xi_{\mathrm{i,k}}, \tag{5.14}$$

where $\xi_{i,k}$ represents the semantic similarity of the ith user over the kth channel, which depends on the neural network structure of DeepSC and channel conditions. It can be expressed as a function of n_i^w and $\gamma_{i,k}$, i.e., $\xi_{i,k} = f(n_i^w, \gamma_{i,k})$. From Eq. (5.14), the corresponding S-SE can be expressed as:

$$\Phi_{i,k} = \frac{\Gamma_{i,k}}{B_W} = \frac{I_{\mathrm{sem}}}{n_i^w O_i} \xi_{\mathrm{i,k}}. \tag{5.15}$$

(2) Problem Formulation

In this part, a semantic-aware resource allocation model is proposed to maximize the overall S-SE of all users. By denoting Φ as the overall S-SE of all users, we have:

$$\Phi = \sum_{i=1}^{N_U} \sum_{k=1}^{N_K} \alpha_{i,k} \frac{\xi_{i,k} I_{\mathrm{sem}}}{n_i^w O_i}. \tag{5.16}$$

The channel assignment vector is considered one of the optimization variables to fully exploit the performance advantage of DeepSC in the low SNR regime. Furthermore, we also optimize the average number of the transmitted semantic symbols for each word, n_i^w, to enable each symbol to carry more semantic information and thus achieve a higher S-SE while ensuring the same transmission reliability.

According to the above analysis, the optimization problem can be formulated as:

$$(\mathbf{P0}) \quad \max_{\alpha_i, n_i^w} \Phi \tag{5.17}$$

$$\text{s.t. } C_1 : \alpha_{i,k} \in \{0, 1\}, \forall n \in \mathcal{U}, \forall k \in \mathcal{K}, \tag{5.17a}$$

$$C_2 : \sum\nolimits_{i=1}^{N_U} \alpha_{i,k} \leq 1, \ \forall k \in \mathcal{K}, \tag{5.17b}$$

$$C_3 : \sum\nolimits_{k=1}^{N_K} \alpha_{i,k} \leq 1, \ \forall n \in \mathcal{U}, \tag{5.17c}$$

$$C_4 : n_i^w \in \{1, 2, \ldots, N^w\}, \tag{5.17d}$$

$$C_5 : \xi_{i,k} \geq \xi_{\mathrm{th}}, \tag{5.17e}$$

$$C_6 : \Phi_{i,k} \geq \Phi_{\mathrm{th}}, \tag{5.17f}$$

where C_1, C_2, and C_3 are channel assignment constraints, C_4 specifies the permitted range of the average number of semantic symbols per word with K representing the maximum value, C_5 reflects the minimum required semantic similarity ξ_{th}, and C_6 restricts the minimum S-SE of users by Φ_{th}.

(3) The Optimal Solution

To solve (**P0**), two challenges need to be addressed. First, we need to handle the term I_{sem}/O_i in the objective function. Second, we need to address $\xi_{i,k}$, which is closely related to Φ, C_5, and C_6.

Regarding the first challenge, we note that the term I_{sem}/O_i is dependent on the source type. According to the analysis in Sect. 5.2.1.2 (1), this term is a constant for a particular source type and does not affect resource optimization. Thus, we can omit this term when solving (**P0**). Consequently, the optimization problem (**P0**) can be rewritten as:

$$\begin{aligned} \textbf{(P1)} \quad & \max_{\boldsymbol{\alpha}_i, n_i^w} \widetilde{\Phi} = \sum_{i=1}^{N_U} \sum_{k=1}^{N_K} \alpha_{i,k} \frac{\xi_{i,k}}{n_i^w} \\ & \text{s.t. } C_1, C_2, C_3, C_4, C_5, C_6. \end{aligned} \tag{5.18}$$

Next, since $\xi_{i,k}$ depends on the specific semantic communication system and the physical channel conditions, we run the DeepSC model over the AWGN channel to obtain the mapping between $\xi_{i,k}$ and $(n_i^w, \gamma_{i,k})$, as shown in Fig. 5.6.

After addressing the two challenges, we can solve (**P0**). Specifically, due to the orthogonality of different cellular links, (**P1**) can be decoupled into the following two equivalent independent optimization problems:

$$\begin{aligned} \textbf{(P2)} \quad & \max_{n_i^w} \widetilde{\Phi}_{i,k} \\ & \text{s.t. } C_4, C_5, C_6, \end{aligned} \tag{5.19}$$

and

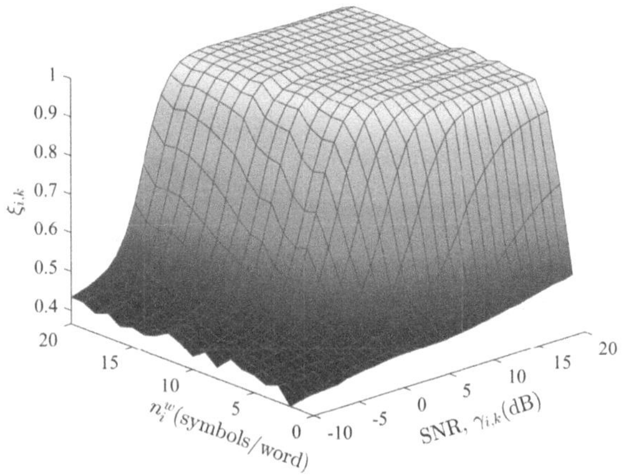

Fig. 5.6 The semantic similarity for DeepSC

$$
\begin{aligned}
(\mathbf{P3}) \quad & \max_{\boldsymbol{\alpha}_i} \sum_{i=1}^{N_U} \sum_{k=1}^{N_K} \alpha_{i,k} \widetilde{\Phi}_{i,k}^{\max} \\
& \text{s.t. } \mathrm{C}_1, \mathrm{C}_2, \mathrm{C}_3,
\end{aligned} \tag{5.20}
$$

where $\widetilde{\Phi}_{i,k} = \xi_{i,k}/n_i^w$ and $\widetilde{\Phi}_{i,k}^{\max}$ represents the maximum $\widetilde{\Phi}_{i,k}$ with respect to n_i^w. (**P2**) targets on obtaining $\widetilde{\Phi}_{i,k}$ for all users over all candidate channels. Since $\xi_{i,k}$ in C_5 and C_6 can only be obtained by the look-up table method, the exhaustive searching method is adopted to solve (**P2**). Moreover, (**P3**) can be regarded as a maximum match problem of a bipartite graph and can be solved using the Hungarian algorithm [8], where two vertex sets are $\mathcal{U}$ and $\mathcal{K}$, and $\widetilde{\Phi}_{i,k}^{\max}$ is regarded as the weight between the ith user and the kth channel.

5.2.1.3 Simulation Results and Comparison

To comprehensively evaluate the performance of the proposed semantic-aware resource allocation scheme comprehensively, we conduct the following verifications in the simulation:

1. Comparing the proposed resource allocation model against the conventional one to verify the proposed model in semantic-aware networks
2. Comparing the S-SE of semantic and conventional communication systems to demonstrate the superiority of semantic communications

Since conventional systems are typically assessed in the bit domain, we first develop a transform method to convert the typical spectral efficiency (SE) to the S-SE, considering the effect of source coding and enabling fair comparisons. Based on this, we present and analyze simulation results.

(1) The Transform Method for Fair Comparisons

In conventional communications, each letter in a word is mapped into bits through a source encoder. From a semantic perspective, each bit can be loosely regarded as a semantic symbol, although it may carry less semantic information than the semantic symbol of DeepSC. Similar to the definition in Sect. 5.2.1.2 (1), the equivalent S-R can be expressed as:

$$
\Gamma'_{i,k} = C_{i,k} \frac{I_{\mathrm{sem}}}{\mu L} \xi_{i,k}, \tag{5.21}
$$

where $C_{i,k}$ is the transmission rate of the ith user over the kth channel, measured in *bits/s*, and μ is defined as the transforming factor that reveals the ability of the source coding scheme in compressing data, representing the average number of bits per word, measured in *bits/word*. For example, if a word includes, on average, five letters, and American standard code for information interchange (ASCII) code is adopted to encode each letter, μ would be 40 bits/word. Additionally, assuming no bit errors in conventional communications, $\xi_{i,k}$ is equal to 1. By denoting $R_{i,k} = C_{i,k}/B_W$ as the SE, the equivalent S-SE can be given by:

$$\Phi'_{i,k} = R_{i,k}\frac{I_{\text{sem}}}{\mu O_i}. \tag{5.22}$$

Hence, by considering both the source coding process and the bit transmission process, we can derive the S-SE of conventional systems, enabling fair comparisons between different communication systems.

(2) Benchmarks

Considering that the proposed resource allocation scheme is specific to the DeepSC semantic system, we compare it with the following three benchmarks, including an ideal system and two widely deployed practical systems:

- *Ideal system*: Shannon limit can be achieved with no bit errors, i.e., $R_{i,k} = \log_2(1+\gamma_{i,k})$.
- *4G system*: According to the measured SNR, the BS obtains the channel quality indicator (CQI)[6], based on which the achievable SE $R_{i,k}$ can be obtained according to Table 7.2.3-1 in 3GPP TS 36.213.
- *5G system*: Similar to 4G, the BS obtains CQI based on the measured SNR [4], and then determines the achievable SE $R_{i,k}$ according to Table 5.2.2.1-2 in 3GPP TS 38.214.

It is important to note that no scheme can achieve a higher bit transmission rate than the ideal system. However, in this chapter, we focus on the S-SE to evaluate the performance. By adopting the developed transform method, the S-SE optimization problem of the above three benchmarks can be formulated as:

$$\textbf{(P4)} \quad \max_{\boldsymbol{\alpha}_i} \sum_{i=1}^{N_U}\sum_{k=1}^{N_K} \alpha_{i,k}\Phi'^{\Delta}_{i,k} \tag{5.23}$$

$$\text{s.t.} \quad \mathrm{C}_1, \mathrm{C}_2, \mathrm{C}_3,$$

$$\mathrm{C}_7 : \ \Phi'^{\Delta}_{i,k} \geq \Phi_{\text{th}}, \tag{5.23a}$$

where $\Phi'^{\Delta}_{i,k}$ represents the S-SE of the ith user over the kth channel in system Δ, in which $\Delta \in \{\text{Ideal, 4G, 5G}\}$. (**P4**) can be solved using the method introduced in Sect. 5.2.1.2 (3).

(3) Simulation Results

In our simulation, we consider a circular network with a radius of $r = 500$ m where N users are uniformly distributed. Unless specifically stated, the relevant parameters are listed in Table 5.4.

First, we examine the conventional resource allocation model in semantic-aware networks. In this simulation, we apply the optimal channel assignment results of

Table 5.4 Simulation parameters

Parameter	Value
Number of users, N_U	5
Number of channels, N_K	5
Channel bandwidth, B_W	180 KHz
Noise power spectral density, N_0	−174 dBm/Hz
path loss model	128.1 + 37.6lg[d(km)] dB
Shadow effect factor	6 dB
Transmit power, p_n	10 dBm
Maximum number of symbols per word, N^w	20 symbols/word
Semantic similarity threshold, ξ_{th}	0.9
S-SE threshold, Φ_{th}	$0.025(I_{sem}/O_i)$ suts/s/Hz
Transforming factor, μ	40 bits/word

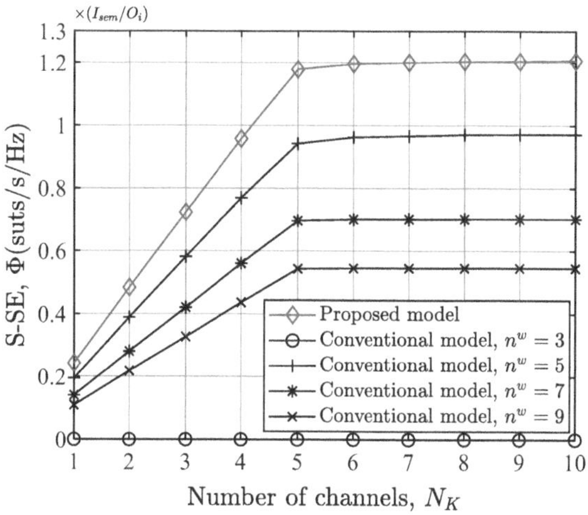

Fig. 5.7 The S-SE of the semantic-aware network with different models

the conventional model in the ideal system to the network, along with different values of n_i^w. We then compare the obtained S-SE with that of the proposed model. As shown in Fig. 5.7, regardless of the value of n_i^w, the S-SE of the conventional model is smaller than that of the proposed model, indicating that the conventional model is not suitable for semantic-aware networks. Additionally, the S-SE of the conventional model with $n_i^w = 3$ is equal to 0 because the semantic similarity is less than the threshold in this case.

Next, we compare the different communication systems with their corresponding resource allocation models. Figure 5.8 shows the S-SE of different systems versus the number of channels. As K increases from 1 to 5, the S-SE of all systems rapidly increases due to serving more users. However, when K keeps on increasing from 5 to 10, the S-SE grows slowly instead of remaining stable because more channels become available, and users can choose channels with higher SNR. Moreover,

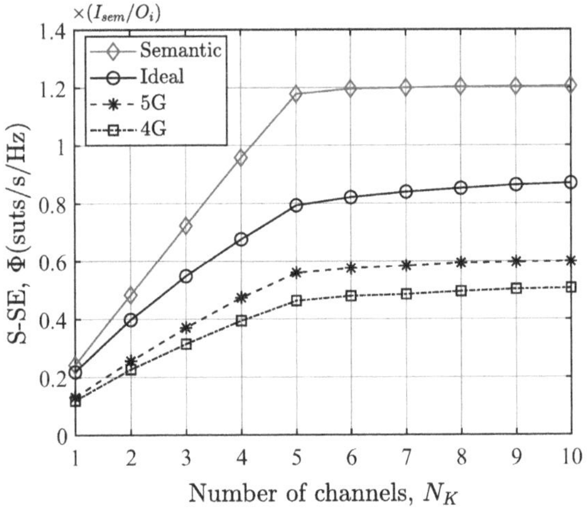

Fig. 5.8 The S-SE versus the number of channels

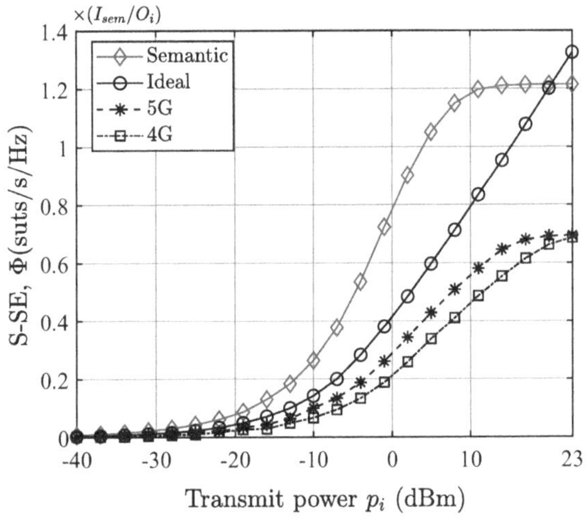

Fig. 5.9 The S-SE versus the transmit power

the semantic communication system outperforms all conventional communication systems.

Figure 5.9 illustrates the S-SE versus the transmit power. As p_i increases, the S-SE of the ideal system increases rapidly, while that of the semantic communication system, 4G system, and 5G system increase first and then tend to reach a constant value. This implies that all practical systems have an upper bound in terms of S-SE with increasing SNR. Furthermore, the semantic communication system shows a larger upper bound than the 4G and 5G systems due to its stronger ability to compress data.

Figure 5.10 shows the S-SE versus the transforming factor. The performance of the semantic communication system remains stable since the transforming factor is irrelevant to it. For the conventional systems, the S-SE decreases with increasing

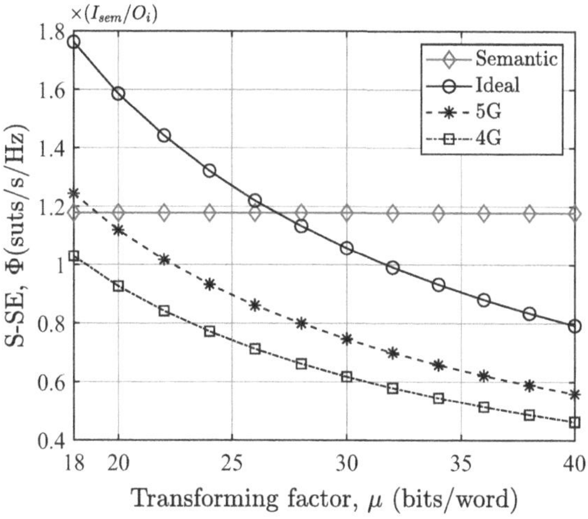

Fig. 5.10 The S-SE versus the transforming factor

μ because S-SE is the ratio of the SE to μ, and the maximum SE is a fixed value with different μ. Additionally, the semantic communication system achieves better performance than both the 4G and 5G systems when μ is larger than 19 bits/word. However, when μ is smaller than approximately 27 bits/word, i.e., when a word can be encoded to less than 27 bits, the semantic communication system performs worse than the ideal system. This figure demonstrates that whether semantic communication systems outperform conventional ones to a great extent depends on the source coding scheme adopted in conventional systems.

5.2.2 *Resource Allocation for a Multitask Network*

5.2.2.1 System Model

In this section, we consider an uplink cellular network with N_B coordinated cells, each containing a BS with N_r receiving antennas and a set of single-antenna users. Let $\mathcal{U}^b$ denote the user set in the bth cell. We assume that N_B BSs form a cooperation cluster, which can exchange CSI and scheduling information with a central controller through high-speed low-delay fiber backhaul links [12]. Additionally, we consider the coexistence of different intelligence tasks in the network, including single-modal tasks and multimodal tasks. Users and BSs are equipped with task-specific semantic communication transmitters and receivers, respectively. The transmission model and semantic communication models are introduced below.

(1) Transmission Model

To fully exploit spectrum resources, we assume that the N_B-coordinated cells share the same channel set denoted by $\mathcal{K}$, and each user can occupy at most one channel with bandwidth B_W. In each cell, the channels are orthogonally allocated to users to eliminate intra-cell interference. With the maximal ratio combining (MRC) detection at the BS, the receiving signal-to-interference-plus-noise ratio (SINR) of user i, $i \in \mathcal{U}^b$ at the bth BS can be expressed as:

$$\gamma_i^b = \sum_{k\in\mathcal{K}} \frac{\alpha_{u,k} p_i |\mathbf{W}_{i,k}^b \mathbf{H}_{i,k}^b|^2}{\|\mathbf{W}_{i,k}^b\|^2\sigma^2 + I_{i,k}^b}, \tag{5.24}$$

where for $\alpha_{i,k} \in \{0, 1\}$, $\alpha_{i,k} = 1$ if channel $k \in \mathcal{K}$ is allocated to user u and $\alpha_{i,k} = 0$ otherwise, p_i is the transmit power of user u, and $\mathbf{H}_{i,k}^b$ is the $N_r \times 1$ channel matrix from user u to the bth BS over channel m. The MRC detection matrix can be obtained by $\mathbf{W}_{i,k}^b = (\mathbf{H}_{i,k}^b)^{\mathrm{H}}$. In Eq. (5.24), σ^2 is the noise power spectral density and $I_{i,k}^b$ is the interference experienced by user u over channel k from the adjacent cells, that is,

$$I_{\mathrm{sem}\,i,k}^{\;b} = \sum_{b'\in\mathcal{B}, b'\neq b} \sum_{i'\in\mathcal{U}^{b'}} \alpha_{i',k} p_{i'} |\mathbf{W}_{i,k}^b \mathbf{H}_{i',k}^b|^2. \tag{5.25}$$

(2) Semantic Communication Models

We focus on two types of intelligence tasks in this chapter: single-modal tasks and bimodal tasks. However, the proposed algorithm can be extended to the case of multiple multimodal intelligence tasks easily. We assume that N_{Si} single-modal users and N_{Bi} bimodal user pairs are randomly deployed in the considered network, with the bth cell having N_{Si}^b single-modal users and N_{Bi}^b bimodal user pairs. For ease of clarification, we consider a single-modal user or a bimodal user pair as a user group. The index set of all user groups in the bth cell is given by:

$$\mathcal{Q}^b = \{1, 2, \ldots, q, \ldots, |\mathcal{Q}^b|\} = \{1, 2, \ldots, N_{\mathrm{Bi}}^b, N_{\mathrm{Bi}}^b{+}1, \ldots, N_{\mathrm{Bi}}^b{+}N_{\mathrm{Si}}^b\}, \tag{5.26}$$

where $|\mathcal{Q}^b| = N_{\mathrm{Bi}}^b + N_{\mathrm{Si}}^b$. Accordingly, the qth user group contains a bimodal user pair when $q \leq N_{\mathrm{Bi}}^b$, and it contains a single-modal user when $q > N_{\mathrm{Bi}}^b$. We denote the set of all user groups in the bth cell as $\mathcal{G}^b = \{\mathcal{G}_q^b\}_{q\in\mathcal{Q}^b}$, where $\mathcal{G}_q^b \subseteq \mathcal{U}^b$ represents the user set of the qth group.

For the single-modal task, we take the text transmission task as an example. Specifically, we adopt DeepSC [22], where the sentence generated by the user is first mapped to semantic symbols that can be directly transmitted over the physical channel by the DeepSC transmitter and then recovered by the DeepSC receiver at the BS. Moreover, semantic similarity is used to evaluate the performance, which is a function of the number of transmitted semantic symbols and SINR. Thus, the semantic similarity of the user group $\mathcal{G}_q^b$, $q > N_{\mathrm{Bi}}^b$ can be expressed as:

$$\xi_q^b = f_{\mathrm{Si}}(n_i^w, \gamma_i^b), u \in \mathcal{G}_q^b, \tag{5.27}$$

where n_i^w is the average number of transmitted semantic symbols of user i.

For the bimodal task, we take the VQA task as an example and adopt the developed DeepSC-VQA model [23]. This task involves two users for text and image transmission. The two users first extract the semantic symbols from the text and image information through the DeepSC-VQA transmitter and then send them to the BS. The received semantic symbols of text and image will be fused by the DeepSC-VQA receiver at the BS to predict the answer. As the two users jointly decide the task performance, the answer accuracy of this task could be modeled as a function with respect to the numbers of transmitted semantic symbols and the SINR of the two users, i.e.,

$$\xi_q^b = f_{\mathrm{Bi}}(n_j^w, n_i^w, \gamma_j^b, \gamma_i^b),\ j, i \in \mathcal{G}_q^b, \tag{5.28}$$

where j and i represent the user for text transmission and the user for image transmission, respectively.

5.2.2.2 Resource Allocation Problem

In this section, we first develop a method to obtain an approximate measure of semantic entropy, based on which a novel QoE model for semantic communications is proposed. Then, the semantic resource allocation is formulated as a QoE maximization problem.

(1) Approximate Measure of Semantic Entropy

Information entropy measures the information based on the statistical characteristics of source symbols, while semantic entropy directly quantifies the semantic information of the source and plays a significant role in evaluating semantic communication efficiency and resource allocation. However, a common semantic information metric is still missing, and most existing metrics are non-calculable [15].

Since the semantic information carried by a source depends on the specific task, semantic entropy should be a measure with respect to the source and the task. Following Chattopadhyay et al. [3], we define the semantic entropy as follows:

Definition 5.1 Given semantic source $\mathcal{X}$, semantic entropy is defined as the minimum expected number of semantic symbols about the data $X \in \mathcal{X}$ that are sufficient to predict task Y, i.e.,

$$\begin{aligned} H(X;Y) &\triangleq \min_{E_\mathrm{S}} \mathbb{E}\left(\dim(\mathrm{Code}^{E_\mathrm{S}}(X))\right) \\ \text{s.t.}\quad & P(Y|\mathrm{Code}^{E_\mathrm{S}}(X)) = P(Y|X), \end{aligned} \tag{5.29}$$

where $\mathrm{Code}^{E_\mathrm{S}}(X)$ denotes the semantic symbol vector extracted from X with the semantic encoder E_S, and $P(Y|X)$ is the conditional probability of Y given X.

From Definition 5.1, the semantic entropy of X given Y is actually defined as an expected value over the whole dataset $\mathcal{X}$, i.e., the semantic entropy is a constant for the same task and dataset. However, it is intractable to find an optimal E_{S} to derive the semantic entropy. To obtain a measure that is both meaningful and manipulable, we utilize a well-designed DL model as the semantic encoder to obtain the approximate semantic entropy for a task, which can be expressed as:

$$\begin{aligned} \tilde{H}(X;Y) &\triangleq \min \mathbb{E}\left(\dim(\mathrm{Code}^{E_{\mathrm{DL}}}(X))\right) \\ \text{s.t.} \quad & P(Y|X) - P(Y|\mathrm{Code}^{E_{\mathrm{DL}}}(X)) < \varepsilon, \end{aligned} \tag{5.30}$$

where the constraint indicates that the gap between $P(Y|X)$ and $P(Y|\mathrm{Code}^{E_{\mathrm{DL}}}(X))$ cannot exceed ε.

According to the aforementioned method, the approximate semantic entropy of the considered tasks can be derived based on the corresponding DL models offline. In particular, we first remove the channel models from DeepSC and DeepSC-VQA, then train them under different settings of the number of semantic symbols, and finally find the minimum number of semantic symbols that can guarantee a performance very close to the upper bound. Furthermore, we define the unit of the approximate semantic entropy as *sut*, as in Sect. 5.2.1.

(2) QoE Model for Semantic Communications

We formulate the QoE model based on two objective metrics: *semantic fidelity* and *semantic rate*. The former represents the accuracy of message transmission, corresponding to the semantic similarity in the considered single-modal task and the answer accuracy in the considered bimodal task. The latter is defined as the amount of semantic information emitted to the transmission medium per second, measured in *suts/s*. The semantic rate of the single-modal user $u \in \mathcal{G}_q^b, q > N_{\mathrm{Bi}}^b$ is:

$$\varphi_i^{\mathrm{sem}} = \frac{\tilde{H}_{\mathrm{Si}}}{n_i^w / B_W}, \tag{5.31}$$

where $\tilde{H}_{\mathrm{Si}}$ represents the DeepSC [22]-based approximate semantic entropy. The semantic rate of bimodal users j and i, $j, i \in \mathcal{G}_q^b, q \leq N_{\mathrm{Bi}}^b$ is expressed as:

$$\varphi_j^{\mathrm{sem}} = \frac{\tilde{H}_{\mathrm{Bi,t}}}{n_j^w / B_W}, \tag{5.32}$$

and

$$\varphi_i^{\mathrm{sem}} = \frac{\tilde{H}_{\mathrm{Bi,i}}}{n_i^w / B_W}, \tag{5.33}$$

respectively, where $\tilde{H}_{\mathrm{Bi,j}}$ and $\tilde{H}_{\mathrm{Bi,i}}$ represent the DeepSC-VQA [23]-based approximate semantic entropy for the text transmission user and the image transmission user, respectively.

It is noteworthy the semantic rate is significantly different from the S-R that is defined as the amount of *successfully* delivered semantic information per second in Sect. 5.2.1. If we denote the S-R as Γ_i, then $\Gamma_i = \varphi_i^{\mathrm{sem}} \xi_q$. The difference between them lies in whether the semantic accuracy is considered. The S-R couples semantic fidelity and semantic rate. However, from the perspective of users, the accuracy and efficiency of message transmission are different, and users may have their own preferences on them depending on the applications. For example, some users prefer high accuracy but are delay-tolerant, while others may desire a higher semantic rate but do not need a very high accuracy. To better reflect the QoE requirements of users, we consider both parameters in formulating the QoE model as:

$$\begin{aligned} QoE_q^b &= \sum_{u \in \mathcal{G}_q^b} w_i G_i^{\mathrm{R}} + (1 - w_i) G_i^{\mathrm{A}} \\ &= \sum_{u \in \mathcal{G}_q^b} \frac{w_i}{1 + e^{\alpha_i (\varphi_i^{\mathrm{req}} - \varphi_i^{\mathrm{sem}})}} + \frac{(1 - w_i)}{1 + e^{\lambda_i (\xi_i^{\mathrm{req}} - \xi_q^b)}}, \end{aligned} \tag{5.34}$$

where QoE_q^b is the QoE of the qth user group in the bth cell, w_i and $(1 - w_i)$ are the weights of semantic rate and semantic accuracy at user u, respectively, G_i^{R} and G_i^{A} are the scores of semantic rate and semantic fidelity at user u, respectively, and α_i and λ_i represent the growth rates of G_i^{R} and G_i^{A}, respectively. Additionally, φ_i^{req} and ξ_i^{req} represent the minimum semantic rate and semantic fidelity required to acquire the 50% of the scores, respectively. Due to the simplicity and generality, we choose the logistic function to model the correlation between QoE and QoS metrics while other QoE function types are applicable as well for the proposed method, such as mean opinion score (MOS)-based function or exponential function. Here, QoE_q^b, G_i^{R}, and G_i^{A} are values between 0 and 1.

(3) Problem Formulation

In this part, we frame the semantic-aware resource allocation problem with the objective of maximizing the overall QoE of all users. This problem is defined in terms of channel assignment, power allocation, and the number of transmitted semantic symbols. The optimization problem can be expressed as follows:

$$\textbf{(P0)} \quad \max_{\{\{\alpha_{i,k}\}, \{p_i\}, \{k_i\}\}} \sum_{b \in \mathcal{B}} \sum_{q \in \mathcal{Q}^b} QoE_q^b \tag{5.35}$$

$$\text{s.t. } \mathrm{C}_1: \alpha_{i,k} \in \{0, 1\}, \ \forall i \in \mathcal{U}^b, \forall k \in \mathcal{K}, \ \forall b \in \mathcal{B}, \tag{5.35a}$$

$$\mathrm{C}_2: \sum_{i \in \mathcal{U}^b} \alpha_{i,k} \le 1, \ \forall k \in \mathcal{K}, \ \forall b \in \mathcal{B}, \tag{5.35b}$$

$$\mathrm{C}_3: \sum_{k\in\mathcal{K}} \alpha_{i,k} \le 1,\ \forall i \in \mathcal{U}^b,\ \forall b \in \mathcal{B}, \tag{5.35c}$$

$$\mathrm{C}_4: \sum_{i\in\mathcal{G}_q^b, q\le N_{\mathrm{Bi}}^b} \alpha_{i,k} \in \{0,2\},\ \forall k \in \mathcal{K},\ \forall b \in \mathcal{B}, \tag{5.35d}$$

$$\mathrm{C}_5: n_i^w \in \mathcal{K}_i,\ \forall i \in \mathcal{U}^b,\ \forall b \in \mathcal{B}, \tag{5.35e}$$

$$\mathrm{C}_6: 0 \le p_i \le P_{\max},\ \forall i \in \mathcal{U}^b,\ \forall b \in \mathcal{B}, \tag{5.35f}$$

$$\mathrm{C}_7: G_i^{\mathrm{R}}, G_i^{\mathrm{A}} \ge G_{\mathrm{th}},\ \forall i \in \mathcal{U}^b,\ \forall b \in \mathcal{B}, \tag{5.35g}$$

where C_1 constrains the range of $\alpha_{i,k}$, C_2 ensures the orthogonal channels for each user in a cell, C_3 restricts each user to occupy at most one channel, C_4 ensures that a bimodal user pair will be allocated no channel or two channels, as assuming only one channel assigned to one of them will lead to a failure, C_5 specifies the range of the number of transmitted semantic symbols for each user, and $\mathcal{K}_i = \mathcal{K}_{\mathrm{Si}}, \mathcal{K}_{\mathrm{Bi,t}}$, and $\mathcal{K}_{\mathrm{Bi,i}}$ when u is the user with the single-modal task, the user with the bimodal task for text transmission, and the user with the bimodal task for image transmission, respectively, C_6 constrains the range of transmit power, and C_7 limits the minimum required scores of semantic rate and semantic accuracy.

5.2.2.3 A Matching Theory-Based Solution

In this section, we first decouple (**P5**) into two separate subproblems and then tackle each one independently to derive a suboptimal solution.

(1) Problem Decoupling
Observation of problem (**P5**) reveals that the optimization variables, $\{\alpha_{i,k}\}$ and $\{p_i\}$, jointly determine the SINR γ_i^b. Additionally, the objective function can be expressed as a function in terms of γ_i^b and n_i^w. Therefore, given γ_i^b, the optimal $\{n_i^w\}$ for each $\mathcal{G}_q^b$ can be obtained by solving the following optimization problem:

$$(\mathbf{P6}) \quad \max_{\{k_i\}, i\in\mathcal{G}_q^b} QoE_q^b \quad \text{s.t. } \mathrm{C}_5 \text{ and } \mathrm{C}_7. \tag{5.36}$$

Then the problems of channel assignment and power allocation problem can be expressed as:

$$(\mathbf{P7}) \quad \max_{\{\{\alpha_{i,k}\},\{p_i\}\}} \sum_{b\in\mathcal{B}} \sum_{q\in Q^b} QoE_q^b \quad \text{s.t. } \mathrm{C}_1, \mathrm{C}_2, \mathrm{C}_3, \mathrm{C}_4, \text{ and } \mathrm{C}_6. \tag{5.37}$$

Hence, for each set of $\{\{\alpha_{i,k}\}, \{p_i\}\}$, the SINR of each user can be calculated using Eq. (5.24). Subsequently, the optimal n_i^w and the corresponding maximum QoE_q^b can be obtained by solving (**P6**). Since ξ_q^b can only be obtained using the lookup table method, the exhaustive search method is utilized to solve (**P6**) in this case. It is worth noting that the complexity remains acceptable due to the limitation of $|\mathcal{K}_i|$. Finally, with the obtained maximum QoE_q^b, we can solve (**P7**) to obtain the solutions, which will be detailed in the next section. The decoupling approach ensures that optimality is not compromised as the optimal solutions of (**P6**) under different SNR values are utilized to solve (**P7**).

(2) Channel Assignment and Power Allocation Subproblem

To address the tight coupling among users in multiple cells and those in a bimodal user pair, we construct a matching game to model (**P7**) and propose a low-complexity matching algorithm to obtain the stable matching in this part.

Problem (**P7**) represents a three-sided many-to-one matching game involving users, channels, and power levels, in which each user can select at most one channel and one power level while each channel or power level can serve multiple users. For ease of algorithm implementation, we combine all combinations of channels and power levels together into a resource set $\mathcal{T} = \{(k, p), \forall k \in \mathcal{K}, \forall p \in \mathcal{P}\}$, where $\mathcal{P}$ denotes the power set. Thus, the three-sided many-to-one matching can be converted into a two-sided matching problem. Furthermore, since the QoE of each user depends not only on the opposite partners they are matched with but also on the users from other cells sharing the same channel, this becomes a matching problem with externalities. The idea of swap matching [1] can be used for reference to obtain the solution.

Considering the two possible cases in each cell, namely, $|\mathcal{U}^b| \leq |\mathcal{K}|$ and $|\mathcal{U}^b| > |\mathcal{K}|$, we focus on optimizing the overall QoE by targeting different market aspects. Specifically, the proposed algorithm aims to increase the utility of users through swap operations to achieve stable matching in the first case, while prioritizing the utility of channels in the second case.

Case 1: $|\mathcal{U}^b| \leq |\mathcal{K}|$ First, we add $|\mathcal{K}| - |\mathcal{U}^b|$ virtual single-modal users to ensure that each channel can be matched with a user, thus facilitating the swap operation. Consequently, the updated user group set in the bth cell becomes $\mathcal{G}_{\mathrm{A}}^b = \mathcal{G}^b \cup \mathcal{U}_0^b$, where $\mathcal{U}_0^b$ represents the set of established virtual single-modal users with $|\mathcal{U}_0^b| = |\mathcal{K}| - |\mathcal{U}^b|$. Correspondingly, we denote the updated index set of $\mathcal{G}_{\mathrm{A}}^b$ as $\mathcal{Q}_{\mathrm{A}}^b$.

As the users within a user group collectively determine the task performance, a bimodal user pair (*rather than an individual bimodal user*) or a single-modal user is treated as a selfish and rational player when making decisions. Thus, there are a total of $|\mathcal{G}_{\mathrm{A}}^b|$ players in the bth cell, and each of them is denoted by $\mathcal{G}_q^b \in \mathcal{G}_{\mathrm{A}}^b$. Clearly, the resource sets of the players with the single-modal tasks and those with the bimodal tasks differ, denoted as $\mathcal{T}_{\mathrm{Si}} = \mathcal{T}$ and $\mathcal{T}_{\mathrm{Bi}} = \{(k, k', p, p'), \forall k, k' \in \mathcal{K}, k \neq k', \forall p, p' \in \mathcal{P}\}$, respectively. Therefore, we represent the resource set of player $\mathcal{G}_q^b$ as $\mathcal{T}_q^b$, where $\mathcal{T}_q^b = \mathcal{T}_{\mathrm{Bi}}$ when $q \leq N_{\mathrm{Bi}}^b$, and $\mathcal{T}_q^b = \mathcal{T}_{\mathrm{Si}}$, otherwise.

Then, we can define a matching Φ as a function mapping from the set $\mathcal{G}_{\mathrm{A}} \cup \mathcal{T}_{\mathrm{A}}$ to the set $\mathcal{G}_{\mathrm{A}} \cup \mathcal{T}_{\mathrm{A}}$, where $\mathcal{G}_{\mathrm{A}} = \bigcup_{b \in \mathcal{B}} \mathcal{G}_{\mathrm{A}}^b$ and $\mathcal{T}_{\mathrm{A}} = \mathcal{T}_{\mathrm{Si}} \cup \mathcal{T}_{\mathrm{Bi}}$. For a matching pair $(\mathcal{G}_q^b, t)$, where $t \in \mathcal{T}_{\mathrm{A}}$, we have $\Phi(\mathcal{G}_q^b) = t$ and $\Phi(t) = \mathcal{G}_q^b$. The utility function of player $\mathcal{G}_q^b$ under matching Φ is defined as:

$$U_q^b(\Phi) = \begin{cases} QoE_q^b, & \text{if } q \in \mathcal{Q}^b; \\ 0, & \text{otherwise.} \end{cases} \tag{5.38}$$

Based on this, whether the considered matching is stable can be judged from the following definition:

Definition 5.2 A matching Φ is stable if and only if, for each player $\mathcal{G}_q^b \in \mathcal{G}_{\mathrm{A}}^b, \forall b \in \mathcal{B}$ with $\Phi(\mathcal{G}_q^b) = t$, there does not exist a blocking matching Φ_q^{B} such that $\exists t' \in \mathcal{T}_q^b$:

1. $\forall i \in \mathcal{G}_{(t)} \cup \mathcal{G}_{(t')}, U_i^b(\Phi_q^{(t,t')}) \geq U_i^b(\Phi_q^{\mathrm{B}})$ and
2. $\exists i \in \mathcal{G}_{(t)} \cup \mathcal{G}_{(t')}, U_i^b(\Phi_q^{(t,t')}) > U_i^b(\Phi_q^{\mathrm{B}})$,

where $\mathcal{G}_{(t)}$ consists of all players who share at least one same channel with those included in t, $\Phi_q^{(t,t')}$ represents the swap matching in which the player $\mathcal{G}_q^b$ swaps t with t', and the players who are in both $\mathcal{G}_{(t')}$ and the bth cell swap their channels accordingly, while the remaining players remain unchanged.

Case 2: $|\mathcal{U}^b| > |\mathcal{K}|$ In this case, if we follow the solution for **Case 1**, no channel will be tentatively matched with $|\mathcal{U}^b| - |\mathcal{K}|$ of all users, even though the overall QoE would increase. To address this issue, we focus on the channel market to achieve stable matching. However, when a channel is matched with a user with the bimodal task, its utility depends not only on the user it matches but also on the channel that the other user in the same group is matched with. Hence, we propose that the user groups perform the swap operation, but the channels make the swap decision based on their utility changes.

Similar to **Case 1**, we first add $|\mathcal{U}^b| - |\mathcal{K}|$ virtual channels to facilitate swap operations at each user. Subsequently, the channel set in the bth cell is updated as $\mathcal{K}_{\mathrm{A}}^b = \mathcal{K} \cup \mathcal{K}_0^b$, where $\mathcal{K}_0^b$ represents the set of established virtual channels with $|\mathcal{K}_0^b| = |\mathcal{U}^b| - |\mathcal{K}|$. Then, the resource sets are redefined in a similar manner but based on $\mathcal{K}_{\mathrm{A}}^b$. For simplicity, we use the same notation $\mathcal{T}_q^b$ to represent the resource set of each player $\mathcal{G}_q^b \in \mathcal{G}^b$. Furthermore, the utility function of channel $k \in \mathcal{K}_{\mathrm{A}}^b$ in the $b$the cell under a matching Φ is defined as:

$$U_m^b(\Phi) = \begin{cases} 0, & \text{if } k \in \mathcal{K}_0^b; \\ U_{\mu(k)}^b(\Phi)/2, & \text{if } k \in \mathcal{K}^b, \mu(k) \leq N_{\mathrm{Bi}}^b; \\ U_{\mu(k)}^b(\Phi), & \text{if } k \in \mathcal{K}^b, \mu(k) > N_{\mathrm{Bi}}^b, \end{cases} \tag{5.39}$$

where $\mu(k) \in Q^b$ represents the index of the player matched with channel k. When the player is a bimodal user pair, $U_m^b(\Phi)$ is set as half the utility of the player, while for a single-modal user, $U_m^b(\Phi)$ is equal to the player's utility. We provide the following definition:

Definition 5.3 A matching Φ is stable if and only if, for each channel $k \in \mathcal{K}_{\mathrm{A}}^b$ with $\mu(k) \in Q^b$, there does not exist a blocking matching Φ_k^{B} such that $\exists t' \in \mathcal{T}_{\mu(k)}^b$:

1. $\forall i \in \mathcal{K}_{(\mathcal{G})}, U_i^b(\Phi_{\mu(k)}^{(t,t')}) \geq U_i^b(\Phi_k^{\mathrm{B}})$ and
2. $\exists i \in \mathcal{K}_{(\mathcal{G})}, U_i^b(\Phi_{\mu(k)}^{(t,t')}) > U_i^b(\Phi_k^{\mathrm{B}})$,

where $\mathcal{K}_{(\mathcal{G})}$ consists of all channels of all players in $\mathcal{G}$, and $\mathcal{G} = \mathcal{G}_{(t)} \cup \mathcal{G}_{(t')}$.

Based on the two definitions, we have developed an efficient matching algorithm to obtain the stable matching, as shown in Algorithm 2. The algorithm starts with a random initial matching, and then users search their resource sets to find any blocking matching and update the current matching accordingly. The algorithm continues until no blocking matching is found, and the stable matching Φ is then the output.

Algorithm 2: Proposed matching algorithm for channel assignment and power allocation

---

Input: $\mathcal{B}$, $\mathcal{U}^b$, $\mathcal{K}_{\mathrm{A}}^b$, $\mathcal{G}_{\mathrm{A}}^b$, $\mathcal{T}_q^b$, $\forall q \in Q_{\mathrm{A}}^b$, $\forall b \in \mathcal{B}$.

```
Initialization: Initialize the matching as Φ, where the matching channels and power levels
    of all users are set as a permutation of all channels and the minimum power level,
    respectively.
repeat
    for all b ∈ B do
        for all G_q^b ∈ G_A^b do
            t = Φ(G_q^b);
            for all t' ∈ T_q^b, t' ≠ t do
                if |U^b| ≤ |K| then
                    If Φ is a blocking matching according to Definition 5.2, update the
                    matching as Φ = Φ_q^(t,t'); otherwise, keep the current matching
                    state.
                else
                    If Φ is a blocking matching according to Definition 5.3, update the
                    matching as Φ = Φ_μ(k)^(t,t') where k is the channel in t; otherwise,
                    keep the current matching state.
until No blocking matching is found;
```

Output: the stable matching Φ.

Remark 5.1 The worst-case complexity of Algorithm 2 is as $O(V \times |\mathcal{B}| \times ((|\mathcal{G}_{\mathrm{A}}^{b}| - N_{\mathrm{Bi}}^{b}) \times |\mathcal{K}_{\mathrm{A}}^{b}| \times |\mathcal{P}| + N_{\mathrm{Bi}}^{b} \times \binom{2}{|\mathcal{K}_{\mathrm{A}}^{b}|} \times |\mathcal{P}|^2))$, where V is the number of iterations. However, the complexity of the exhaustive search method here is as $O((\frac{\max(|\mathcal{U}^b|, K)!}{\min(|\mathcal{U}^b|, K)!} \times |\mathcal{P}|^{|\mathcal{U}^b|})^{|\mathcal{B}|})$.

Remark 5.2 Since the utility of user groups or channels is bounded by 1 and increases monotonically through swap operations, Algorithm 2 will terminate with a final matching after a finite number of iterations.

5.2.2.4 Simulation Results

In this section, we present simulation results to demonstrate the performance of the proposed QoE-aware resource allocation method in a multicell multitask network.

We consider three cells in the simulation, where users are randomly deployed. The radius of each cell is set to 500 m. Each BS is equipped with two receiving antennas, while each user has a single antenna. The channel model takes into account both large-scale fading, including path loss and shadowing, and small-scale Rayleigh fading. The adopted path loss model is $128.1 + 37.6\lg[\mathrm{d(km)}]$ dB and the shadowing factor is set to 6 dB. The bandwidth of each channel is $B_W = 180$ kHz. The noise power spectral density is $\sigma^2 = -174\,\mathrm{dBm/Hz}$. In the simulation, we consider the available power levels as $\mathcal{P} = \{-10, -5, 0, 5, 10, 15, 20\}\,\mathrm{dBm}$. Additionally, each user generates its QoE-related parameters as follows: $w_i \sim U(0, 1)$, $\xi_i^{\mathrm{req}} \sim U(0.8, 0.9)$, and $\lambda_i \sim N(55, 2.5^2)$. For text transmission users, we set $\varphi_i^{\mathrm{req}} \sim U(50, 70)$ in ksuts/s and $\alpha_i \sim N(0.2, 0.05^2)$. For image transmission users, we set $\varphi_i^{\mathrm{req}} \sim U(80, 100)$ in ksuts/s and $\beta_i \sim N(0.1, 0.02^2)$. The threshold for the scores is set as $G_{\mathrm{th}} = 0.5$. Furthermore, we run the DeepSC model and DeepSC-VQA model to obtain the mapping relations $\xi_q^b = f_{\mathrm{Si}}(n_i^w, \gamma_u^b)$ and $\xi_q^b = f_{\mathrm{Bi}}(k_{ut}, k_{ui}, \gamma_{ut}^b, \gamma_{ui}^b)$ based on the settings as $\mathcal{K}_{\mathrm{Si}} = \{1, 2, \ldots, 20\}$, $\mathcal{K}_{\mathrm{Bi,t}} = \{2, 4, 6, 8, 10\}$, and $\mathcal{K}_{\mathrm{Bi,i}} = \{394, 788, 1576, 2364, 3152\}$, respectively.

First, we verify the superiority of the developed QoE-based formulation. Figure 5.11 compares the QoE maximization and S-R maximization methods presented in Sect. 5.2.1. The upper bound is obtained by assuming that the maximum number of users in each cell, i.e., $\max(|\mathcal{U}^b|, |\mathcal{K}|)$, can be served and their QoE can reach 1. Thus, the upper bound is independent of G_{th} and is used to demonstrate the effectiveness of the proposed method. As G_{th} increases, the overall QoE of both methods decreases due to the tighter restriction. However, the overall QoE of the proposed method remains 13, while the S-R maximization method shows worse performance, especially when $G_{\mathrm{th}} > 0.5$. This implies that the proposed method can better adapt to the changes in user requirements.

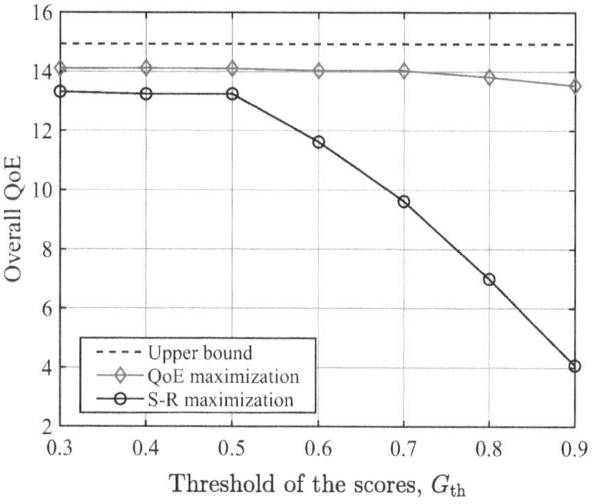

Fig. 5.11 The overall QoE of the proposed QoE maximization method and the S-R maximization method with $(N_{Si}, N_{Bi}) = (6, 6)$ and $|\mathcal{K}| = 6$

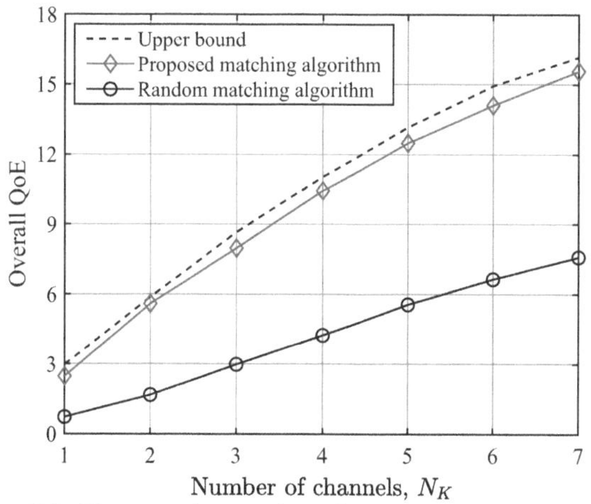

Fig. 5.12 The overall QoE of different algorithms with $(N_{Si}, N_{Bi}) = (6, 6)$

Figure 5.12 illustrates the overall QoE versus the number of channels for different algorithms. Due to the unacceptable computational complexity of the exhaustive search method, as analyzed in Sect. 5.2.2.3 (2), we do not compare the proposed matching algorithm with it. Instead, we compare it with the upper bound of overall QoE, along with the random matching method. From this figure, it is evident that the proposed matching algorithm outperforms the random method significantly and closely approaches the upper bound.

Finally, we compare the proposed method with and without multicell cooperation in Fig. 5.13. The method with multicell cooperation achieves better performance than the one without, which verifies the effectiveness of the proposed method in coping with inter-cell interference.

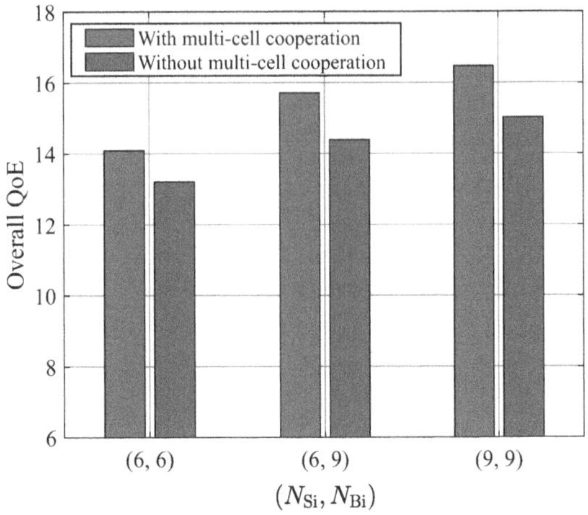

Fig. 5.13 The overall QoE of the proposed method with and without multicell cooperation with $|\mathcal{K}| = 6$

References

1. Bodine-Baron, E., Lee, C., Chong, A., Hassibi, B., Wierman, A.: Peer effects and stability in matching markets. In: Proceedings of the International Symposium on Algorithmic Game Theory, pp. 117–129. Springer, Berlin (2011)
2. Bourtsoulatze, E., et al.: Deep joint source-channel coding for wireless image transmission. IEEE Trans. Cogn. Commun. Netw. **5**(3), 567–579 (2019)
3. Chattopadhyay, A., Haeffele, B.D., Geman, D., Vidal, R.: Quantifying task complexity through generalized information measures. https://openreview.net/pdf?id=vcKVhY7AZqK (2020). Accessed 24th Dec 2021
4. Chu, E., Yoon, J., Jung, B.C.: A novel link-to-system mapping technique based on machine learning for 5G/IoT wireless networks. Sensors **19**(5), 1196 (2019)
5. Gao, J., et al.: A survey on deep learning for multimodal data fusion. Neur. Comput. **32**(5), 829–864 (2020)
6. Ghosh, A., Ratasuk, R.: Essentials of LTE and LTE-A. Cambridge University Press, Cambridge (2011). https://doi.org/10.1017/CBO9780511997082
7. Jankowski, M., Gündüz, D., Mikolajczyk, K.: Wireless image retrieval at the edge. IEEE J. Sel. Areas Commun. **39**(1), 89–100 (2021). https://doi.org/10.1109/JSAC.2020.3036955
8. Kuhn, H.W.: The hungarian method for the assignment problem. Naval Res. Logist. Quart. **2**, 83–97 (1955)
9. Lahat, D., et al.: Multimodal data fusion: an overview of methods, challenges, and prospects. Proc. IEEE **103**(9), 1449–1477 (2015)
10. Lee, C., et al.: Deep learning-constructed joint transmission-recognition for internet of things. IEEE Access **7**, 76547–76561 (2019)
11. Liu, C., Guo, C., Yang, Y., Jiang, N.: Adaptable semantic compression and resource allocation for task-oriented communications (2022). arXiv:2204.08910
12. Lu, X., Ni, Q., Li, W., Zhang, H.: Dynamic user grouping and joint resource allocation with multi-cell cooperation for uplink virtual MIMO systems. IEEE Trans. Wirel. Commun. **16**(6), 3854–3869 (2017). https://doi.org/10.1109/TWC.2017.2689760
13. Nariman, F., et al.: Deep learning for joint source-channel coding of text. In: Proceedings of the IEEE International Conference on Acoustics, Speech, and Signal Processing (ICASSP), pp. 2326–2330, Calgary, AB, Canada (2018)
14. Papineni, K., Roukos, S., Ward, T., Zhu, W.J.: Bleu: A method for automatic evaluation of machine translation. In: Proceedings ofthe Annual Meeting of the Association for Computational Linguistics (ACL). Philadelphia, Pennsylvania (2002)

15. Qin, Z., Tao, X., Lu, J., Li, G.Y.: Semantic communications: Principles and challenges (2021). arXiv: 2201.01389
16. Qin, Z., et al.: Deep learning in physical layer communications. IEEE Wire. Comm. **26**(2), 93–99 (2019)
17. Reimers, N., Gurevych, I.: Sentence-BERT: Sentence embeddings using siamese BERT-networks. In: Proceedings of the Empirical Methods in Natural Language Processing (EMNLP) (2019)
18. Shi, G., et al.: A new communication paradigm: from bit accuracy to semantic fidelity. arXiv preprint arXiv:2101.12649 (2021)
19. Wang, Y., Chen, M., Luo, T., Saad, W., Niyato, D., Poor, H.V., Cui, S.: Performance optimization for semantic communications: an attention-based reinforcement learning approach. IEEE J. Sel. Areas Commun. **40**(9), 2598–2613 (2022). https://doi.org/10.1109/JSAC.2022.3191112
20. Weng, Z., et al.: Semantic communication systems for speech transmission. IEEE J. Sel. Areas Commun. **39**(8), 2434–2444 (2021)
21. Xia, L., Sun, Y., Li, X., Feng, G., Imran, M.A.: Wireless resource management in intelligent semantic communication networks. In: Proceedings of the IEEE Conference on Computer Communications Workshops (INFOCOM WKSHPS), pp. 1–6, New York, USA (2022). https://doi.org/10.1109/INFOCOMWKSHPS54753.2022.9797984
22. Xie, H., Qin, Z., Li, G.Y., Juang, B.H.: Deep learning enabled semantic communication systems. IEEE Trans. Signal Process. **69**, 2663–2675 (2021). https://doi.org/10.1109/TSP.2021.3071210
23. Xie, H., Qin, Z., Tao, X., Letaief, K.B.: Task-oriented multi-user semantic communications. IEEE J. Sel. Areas Commun. **40**(9), 2584–2597 (2022). https://doi.org/10.1109/JSAC.2022.3191326
24. Xie, H., et al.: A lite distributed semantic communication system for internet of things. IEEE J. Sel. Areas Commun. **39**(1), 142–153 (2021)
25. Zuo, J., et al.: Computational complexities and relative performance of ldpc codes and turbo codes. In: Proceedings of the International Conference on Software Engineering and Service Science, pp. 251–254, Beijing (2013)

Chapter 6
QoE Optimization for Wireless Multimedia Communications

6.1 Rate-Semantic-Distortion-Based Image Compression

Recent works have primarily focused on semantic communications for text/speech modalities [41, 54–56], thus falling under the category of *text semantic communications* (TSC). In [18], a deep learning–based joint source-channel coding (JSCC) approach for text is proposed, which achieves a lower word error rate and preserves semantic information of sentences. Later, in [34], the authors design a reinforcement learning (RL)–based semantic communication system to handle the non-differentiability of semantic metrics and interact with the surrounding noisy environment.

However, the visual modality, especially video/image traffic, currently accounts for 75% of IP traffic [15], making it more informative than the text modality. Nevertheless, the transmission of large volumes of multimedia traffic faces challenges such as delay and network congestion, which are difficult to address using conventional communication techniques. Therefore, it is of greater importance to establish the *image semantic communications* (ISC) that can significantly reduce the amount of data to be transmitted without sacrificing the semantic fidelity of the image.

In this chapter, we propose an RL-based adaptive semantic coding (RL-ASC) approach to jointly address the issue of semantic similarity and perceptual performance issues in ISC. By relaxing of precision requirements, we can alleviate the bandwidth burden. Our contributions are summarized as follows:

- We introduce a novel representation unit called a *semantic concept*, which includes the category, features, and spatial relations of each object. A convolutional semantic encoder is proposed to extract semantic concepts at the transmitter and convert them into bitstreams.
- We propose a task-driven image semantic coding framework that utilizes a new reconstruction criterion, namely, *rate-semantic-perceptual* loss. We present an

X. Tao et al., *Wireless Multimedia Computational Communications*, Wireless Networks, https://doi.org/10.1007/978-3-031-64155-8_6

RL-based semantic bit allocation model to determine optimal discrete quantization coefficients and achieve adaptive coding.

- We propose an attention-based generative semantic decoder at the receiver to reconstruct the image with high perceptual quality even at aggressive compression ratios. Specifically, both global and local generators are learned to capture global and category-wise features.

6.1.1 *The RL-Based Adaptive Semantic Coding*

In this section, we propose the semantic concept as the representation unit and adopt semantic fidelity and perceptual quality as the optimization criteria. We then design an RL-based semantic bit allocation model to dynamically highlight and encode the task-related semantic concepts.

The system model of the proposed RL-ASC is illustrated in Fig. 6.1, which consists of three modules: the semantic encoder, the agent π responsible for adaptive quantization, and the semantic decoder. During the inference stage, these modules are pretrained and kept fixed to accommodate a specific downstream task. The semantic encoder extracts the semantic information of each object class from the input image. Then, each class of objects is assigned with an optimal quantization level learned by the RL agent π to ensure an aggressive bit rate Φ. At the receiver, the semantic decoder reconstructs all the semantic information in parallel, achieving good performance in terms of both semantic losses L_s and perceptual loss L_p.

6.1.1.1 Representation Unit: From Pixel to Semantic Concept

Existing image communications typically use pixels as the representation unit and aim to reconstruct the transmitted image exactly as it was transmitted. In contrast,

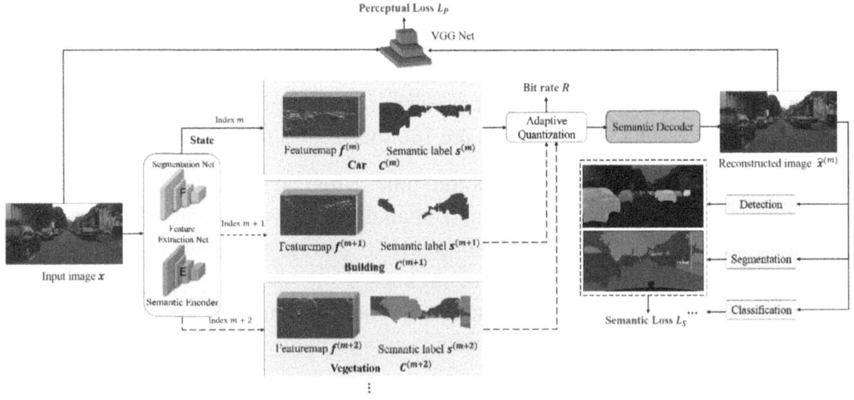

Fig. 6.1 System model of the proposed RL-based adaptive semantic coding method

the ISC interprets the underlying semantics of the image and thus tolerates certain pixel errors. In this study, we focus on the urban street scene scenario and utilize the *Cityscapes* dataset, where the quantity and classes of the semantic concepts are predefined. The dataset contains $D = 30$ number of classes grouped into the following categories: flat surfaces, humans, vehicles, constructions, objects, nature, sky, and void. If the image details do not fit into any of these classes, then it belongs to the void category.

Let $I \in \mathbb{R}^{3\times H_{\text{image}}\times W_{\text{image}}}$ denote the input image, where "3" represents the RGB channels, and H_{image} and W_{image} are the dimensions along the height and width directions, respectively. At the transmitter, we design the semantic encoder at the transmitter to extract semantic information for individual categories. The input image I is fed into the segmentation network $\mathcal{F}$, which generates the semantic label map $\boldsymbol{M}_s = \mathcal{F}(I)$, where $\boldsymbol{M}_s \in \mathbb{R}^{W_{\text{image}}\times H_{\text{image}}}$ assigns a class label to each pixel in I, revealing the spatial arrangement of the M total objects. The entry of $\boldsymbol{M}_s$ at pixel coordinate (x, y) is an integer $\boldsymbol{M}_s(x, y) \in \{1, 2, \ldots, M\}$, representing the class label. We denote $\boldsymbol{M}_s^{(t)}$, where $m = 1, 2, \ldots, M$ is the label ID, as the *semantic mask* of the mth semantic concept $\mathbf{C}^{(t)}$, the entry of which is expressed as:

$$\boldsymbol{M}_s^{(t)}(x, y) = \begin{cases} 1, & \text{if } \boldsymbol{M}_s(x, y) = m, \\ 0, & \text{otherwise.} \end{cases} \tag{6.1}$$

To meet the transmission bit rate requirements, the semantic information needs to be downscaled and then transformed into a bitstream. The downscaled semantic mask is denoted as $\boldsymbol{M}_{s_d}^{(t)} \in \mathbb{R}^{w\times h}$, where $w \leq W_{\text{image}}$ and $h \leq H_{\text{image}}$.

Meanwhile, I is fed into the feature extraction network W_E that convolutionally processes I into a feature map of size $N_K \times w \times h$. This feature map is then projected down to k channels ($k < N_K$) to reduce dimensionality and discard perceptually redundant features, ensuring compliance with the bit rate requirements. The resulting feature map $\boldsymbol{f} \in \mathbb{R}^{k\times w\times h}$ undergoes instance normalization and Leaky ReLU activation after each convolutional layer. The category-specific feature maps $\boldsymbol{f}^{(t)} \in \mathbb{R}^{k\times w\times h}$ for $\mathbf{C}^{(t)}$ are obtained by decomposing $\boldsymbol{f}$:

$$\boldsymbol{f}^{(t)} = \boldsymbol{f} \odot \boldsymbol{M}_{s_d}^{(t)}, \tag{6.2}$$

where $\odot$ represents the element-wise multiplication.

The filtered category-specific feature maps $\boldsymbol{f}^{(t)}$ should possess sufficient discriminative power. We design a novel classification module that employs the feature classification loss L_C to ensure distinctiveness. The architecture of the classification module is as follows: First, all the $\boldsymbol{f}^{(t)}$'s are fed into a max-pooling layer, resulting in M pooled feature maps with dimensions of $k \times 1 \times 1$. Then, a fully connected (FC) layer with shared parameters across different $\boldsymbol{f}^{(t)}$'s is employed, and the outputs are M logits with a dimension of $M \times 1$. The subsequent softmax function generates the predicted classification probability $\hat{p}^{(t)}$ for each category-specific feature map $\boldsymbol{f}^{(t)}$, while the one-hot ground truth label is denoted as $p^{(t)}$. The feature classification loss L_C is defined as the cross-entropy loss:

$$L_C = -\sum_{t=1}^{T} p^{(t)} \log \hat{p}^{(t)}. \tag{6.3}$$

In this way, we mimic the human image understanding process and divide I into M individual semantic concepts, each represented by a set $\mathbf{C}^{(t)} = \{\boldsymbol{f}^{(t)} \in \mathbb{R}^{n \times w \times h}, \boldsymbol{M}_{sd}^{(t)} \in \mathbb{R}^{w \times h}\}$. Therefore, $\mathbf{C}^{(t)}$ can be considered a novel representation unit that is far more effective than a pixel unit.

6.1.1.2 Reconstruction Metric: From Pixel Loss to Semantic-Perceptual Loss

Conventional image communication systems measure pixel loss using metrics such as peak signal-to-noise ratio (PSNR), and the coding process is optimized based on rate-distortion theory. However, the main objective of ISC is to transmit the underlying meaning of an image rather than an exact pixel copy. ISC aims to extract, encode, and reconstruct each semantic concept $\mathbf{C}^{(t)}$ with a low bit rate, ensuring that critical concepts are recovered with high semantic fidelity and perceptual quality. To achieve this, we propose a novel rate-semantic-perceptual criterion to optimize the semantic coding process.

Intelligent analysis tasks, such as object detection, semantic segmentation, pose estimation, and action recognition, accurately extract semantic information. This semantic information can also be utilized in the image coding process. We define semantic loss for different tasks accordingly. Let $\mathcal{H}_{seg}$ represent the semantic segmentation network. The prediction result for the input image I and the reconstructed image $\hat{I}$ is denoted as $\mathcal{H}_{seg}(I)$ and $\mathcal{H}_{seg}(\hat{I})$, respectively. We propose a novel semantic loss L_{seg} metric that measures the precision performance of the intelligent task $\mathcal{H}_{seg}$ on the reconstructed image $\hat{I}$ given the ground truth semantic label of the original image I.

To implement the semantic segmentation task $\mathcal{H}_{seg}$, we can utilize PSPNet [61]. The predicted object bounding set of class m on $\hat{I}$ is denoted as $\mathcal{H}_{seg}(\hat{I}) = \boldsymbol{B}_d^m$, while the ground truth bounding set on I is denoted as $\mathcal{H}_{seg}(I) = \boldsymbol{B}_g^m$. The precision performance of semantic segmentation can be evaluated using the intersection over union (IoU), which measures the ratio between the union area and the intersection area of the ground truth $\boldsymbol{B}_g^m$ and the prediction $\boldsymbol{B}_d^m$:

$$\begin{aligned} IoU(\mathcal{H}_{seg}(I), \mathcal{H}_{seg}(\hat{I})) &= IoU(\boldsymbol{B}_g^m, \boldsymbol{B}_d^m) \\ &= overlap(\boldsymbol{B}_g^m, \boldsymbol{B}_d^m)/union(\boldsymbol{B}_g^m, \boldsymbol{B}_d^m), \end{aligned} \tag{6.4}$$

where IoU is a value between 0 and 1. Mean IoU (mIoU) represents the average segmentation precision over the total M classes. The semantic loss for semantic segmentation is calculated using mIoU and represents the extent of the non-interpretability of the reconstructed image $\hat{I}$:

$$L_{seg} = 1 - \frac{1}{M}\sum_{t=1}^{T} IoU(\boldsymbol{B}_g^m, \boldsymbol{B}_d^m). \tag{6.5}$$

An ideal $\hat{\boldsymbol{x}}$ should preserve the necessary semantic information required by PSPNet, resulting in $L_s = 0$. In this chapter, we consider the semantic segmentation task in the street scene transmission scenario and use L_{seg} as the semantic metric.

However, the evaluation metrics for some semantic analysis tasks are often non-differentiable and cannot be optimized in an end-to-end manner. Existing works often rely on the assumption of a differentiable semantic encoder and channel model, which is not feasible in real wireless scenarios. To address this issue, inspired by Lu et al. [35] and Li et al. [33], we employ deep reinforcement learning to simplify the solution of the complex optimization problem, rather than directly integrating the semantic fidelity metric into the coding framework. Specifically, Lu et al. [35] focuses on text transmission and designs an RL scheme to optimize the BLEU score and inception score derived from the BERT extractor. Building upon this, we further study the image transmission scenario and develop an RL-based semantic loss.

Additionally, visual performance, such as the naturalness and clarity of $\hat{\boldsymbol{x}}$, is another critical aspect for the ISC. The perceptual loss projects the images I and $\hat{I}$ to a high-dimensional feature space using a pretrained model and minimizes the distance in the high-level space to measure the similarity between the two images. The perceptual loss has been validated to properly predict human scores for image degradation and can be expressed as:

$$L_P = \|\phi(I) - \phi(\hat{\boldsymbol{x}})\|_2^2, \tag{6.6}$$

where $\phi(\cdot)$ represents the pretrained visual geometry group (VGG) network that maps an input image to a high dimensional feature space.

We then formulate the rate-semantic-perceptual criterion L as a triple trade-off between semantic fidelity L_{seg}, perceptual quality L_P, and bit rates φ of the semantic concepts:

$$L = \lambda\varphi + L_{seg} + \eta L_P, \tag{6.7}$$

where φ is measured by the length of the transmitted bitstream, and λ and η are the weighting parameters that balance the three terms. Different points on the rate-semantic-perceptual curve can be obtained by adjusting the values of λ and η.

6.1.1.3 The RL-Based Semantic Bit Allocation Model

Different semantic concepts have varying importance in downstream analysis tasks, and only the task-related semantic information needs to be transmitted with high precision. Critical semantic concepts should be focused on and encoded precisely without semantic loss, while the precision requirements for other objects can be relaxed. For example, in the Cityscapes street scene dataset, the object detection task primarily focuses on the vehicles, pedestrians, and traffic signs, while the background elements such as the sky and vegetation do not significantly impact the core meaning of the image. Furthermore, a few bits are sufficient to represent the sky and vegetation, since their texture is simple and regular.

To determine the critical semantic concepts and assign appropriate precision requirements, we propose an RL-based semantic bit allocation model. After all the semantic concepts are obtained by the semantic encoder, the coding process is modeled as a Markov decision process (MDP), where each semantic concept is sequentially encoded and decoded with an appropriate bit rate in the label index order from $m = 1$ to $m = M$. The RL agent dynamically assigns the quantization level for each semantic concept by balancing the bit cost and reconstruction performance. By integrating semantic fidelity and perceptual quality metrics into the semantic coding optimization, we can represent more informative and visually salient objects with higher bit rates, while allocating lower bit rates to irrelevant regions. When the training of the RL agent is completed, the whole process of the quantization decision is off-policy, which can be processed in parallel with the encoder.

The RL framework typically consists of a state space $\mathbb{S}$ which encompasses all possible information that the agent can observe from the environment. In the tth step, the state is defined as $state^{(t)} = \{\boldsymbol{f}^{(t)}, \boldsymbol{M}_s^{(t)}, t\}$, and we denote $\hat{I}^{(t-1)}$ as the reconstructed image from the former step. The action space $\mathbb{A}$ is defined as a set of quantization levels $action^{(t)} \in \{l_1, l_2, \ldots, l_Q\}$ assigned to the semantic concepts, where Q is the total number of levels. A higher quantization level corresponds to a higher bit rate and finer details. The behavior of the agent is defined by a policy $\pi : \mathbb{S} \times \mathbb{A} \rightarrow [0, 1]$, which maps states to a distribution of actions, denoted as $\pi(action^{(t)}|state^{(t)})$. Specifically, π can be implemented using a neural network parameterized by θ. The intermediate reward $r^{(t+1)}$ is received after the agent π taking action $action^{(t)}$ at state $state^{(t)}$ by evaluating the performance of the current reconstructed image $\hat{I}^{(t)}$. The transition from $trans(state^{(t)}, action^{(t)})$ predicts the next state given the current state and the action. In this work, the evolution from state $state^{(t)}$ to $state^{(t+1)}$ is deterministic, and thus $p(state^{(t+1)}) = trans(state^{(t)}, action^{(t)}) = 1$.

As shown in Fig. 6.2, the detailed procedure of the agent π is as follows: The current state $state^{(t)}$ is fed into the agent to learn the policy π, which determines the semantic importance and accordingly produces an action $action^{(t)}$ to adaptively quantize $\boldsymbol{C}^{(t)}$. The semantic decoder takes the received bitstream and the selected quantization level as input to produce the current reconstructed image $\hat{\boldsymbol{x}}^{(t)}$. The immediate reward $r^{(t+1)}$ is defined as the quality improvement from $\hat{I}^{(t)}$ to $\hat{I}^{(t+1)}$ in terms of the increased semantic similarity, perceptual quality, and decreased bit rate φ. Notably, compared to $\hat{\boldsymbol{x}}^{(t-1)}$, the current $\hat{\boldsymbol{x}}^{(t)}$ is only updated in the region corresponding to particular semantic concept $\boldsymbol{C}^{(t)}$, while other regions remain unchanged. The bit rate φ of $\hat{\boldsymbol{x}}^{(t)}$ is denoted as $\psi(\hat{\boldsymbol{x}}^{(t)})$ and calculated as the sum of

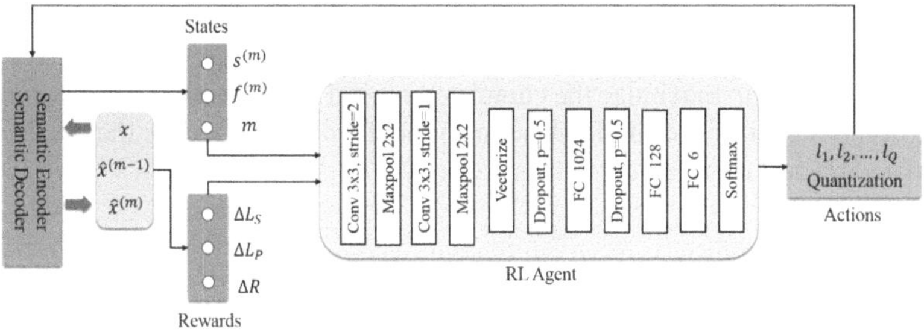

Fig. 6.2 The architecture of the RL-based semantic bit allocation model

bit rates for the processed category–specific features $\boldsymbol{f}^{(t)}$'s and the segmentation map $\boldsymbol{M}_s$. The detailed method for bit rate calculation method is explained in Sect. 6.1.2.

Next, we design the architecture of the RL agent π. The state $state^{(t)}$ is first fed into two consecutive convolutional layers, each of which is followed by the ReLU activation and max-pooling operation. The output of the convolutional layer is then flattened into a vector and passed through into three fully connected (FC) layers to obtain $M \times 1$ dimensional logits. Dropout layer and ReLU activation are employed to prevent over-fitting. Subsequently, the subsequent softmax function generates the predicted probability distribution for each quantization level, and the action $action^{(t)}$ is sampled from this distribution.

At the initialization step, we set the coarsest quantization level l_1 for the entire image resulting in an initial reconstruction $\hat{\boldsymbol{x}}^{(0)}$ with the lowest bit rate $\psi(\hat{\boldsymbol{x}}^{(0)})$. Then, both of the original image I and $\hat{\boldsymbol{x}}^{(0)}$ are fed into the downstream analysis network $\mathcal{H}_{seg}$ (such as PSPNet) and the VGG network to measure the semantic loss and perceptual loss. In the subsequent step, the state $state^{(1)} = \{\boldsymbol{f}^{(1)}, \boldsymbol{M}_s^{(1)}, m = 1\}$ is considered, and the agent generates the action $action^{(1)}$ to select an appropriate quantization level. The current reconstructed image $\hat{I}^{(1)}$ is obtained using the semantic decoder, and the reconstruction performance is sequentially evaluated. At step m, the rate-semantic-perceptual loss is given by:

$$
\begin{aligned}
L^{(t)} &= \lambda[\psi(\hat{\boldsymbol{x}}^{(t)}) - \psi(\hat{\boldsymbol{x}}^{(0)})] \\
&+ [1 - \frac{1}{M}\sum_{t=1}^{T} IoU(\boldsymbol{B}_g^m, \boldsymbol{B}_d^m)] + \eta[\|\phi(I) - \phi(\hat{\boldsymbol{x}}^{(t)})\|_2].
\end{aligned} \tag{6.8}
$$

At step $\tau + 1$, the $(t + 1)$th semantic concept $\mathbf{C}^{(t+1)}$ is updated on the current reconstruction $\hat{I}^{(t+1)}$, and the rate-semantic-perceptual loss $L^{(t+1)}$ is consequently obtained. The intermediate reward is calculated as:

$$
r^{(t+1)} = L^{(t)} - L^{(t+1)} = \lambda\Delta\varphi + \Delta L_S + \eta\Delta L_P, \tag{6.9}
$$

where $\triangle L_S$, $\triangle L_P$, and $\triangle R$ represent the differences in the semantic loss, the perceptual loss, and the bit rate before and after coding.

The agent aims to maximize the cumulated reward over the entire episode, which is the optimal goal of the MDP. Denoting γ as the discount factor, the discounted cumulated reward for a full episode is given by:

$$G = \sum_{t=1}^{T} \gamma^t r^{(t+1)}. \tag{6.10}$$

The objective function for this learning process involves maximizing the expected G across all trajectories. Let T be a complete trajectory $(state^{(1)}, action^{(1)}, r^{(2)}, state^{(2)}, action^{(2)}, r^{(3)}, \ldots, state^{(T)}, action^{(T)}, r^{(T+1)})$, and T_π be the trajectory distribution. Considering the total number of sampled trajectories as N_{traj}. Then the objective function can be written as:

$$J(\pi) = \mathbb{E}_{T_\pi} G = \sum_{N_{\text{traj}}} T_\pi G. \tag{6.11}$$

The optimal policy π can be obtained by applying the gradient ascent method to the sampled trajectories. The derivative of J_π can be calculated as:

$$\nabla_\theta J_\pi = \sum_{N_{\text{traj}}} G \nabla_\theta T_\pi = \sum_{N_{\text{traj}}} G T_\pi \frac{\nabla_\theta T_\pi}{T_\pi} = \sum_{N_{\text{traj}}} T_\pi G \nabla_\theta \log T_\pi. \tag{6.12}$$

In practice, a one-time Monte-Carlo rollout is used to sample a trajectory, leading to the following expression for Eq. (6.12):

$$\nabla_\theta J_\pi = G \sum_{t=1}^{T} \nabla_\theta \log \pi(action^{(t)}|state^{(t)}). \tag{6.13}$$

Then, the parameters of the RL agent π are updated as:

$$\theta \leftarrow \theta + \alpha \nabla_\theta J_\pi, \tag{6.14}$$

where α represents the learning rate.

6.1.2 Semantic Decoder and Training Details

As shown in Fig. 6.3, we discuss the design of the soft quantization, entropy coding, semantic decoder and provide training details in this section.

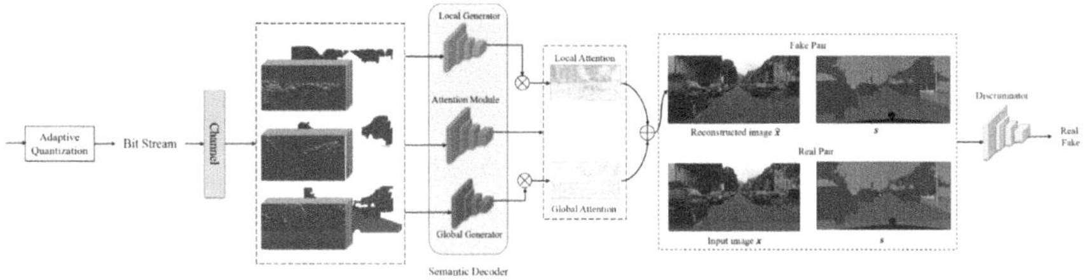

Fig. 6.3 The architecture of the semantic decoder

6.1.2.1 Soft Quantization and Entropy Coding

The RL agent π selects the action $a^{(t)}$ for each $\boldsymbol{f}^{(t)}$ to adaptively assign bits and focus on semantically important regions. We adopt a scalar variant for the quantization approach and obtain the discrete $\hat{\boldsymbol{f}}^{(t)}$ as follows:

$$\hat{\boldsymbol{f}}^{(t)} := Quantize(\boldsymbol{f}^{(t)}, a^{(t)}), \tag{6.15}$$

where each entry of $\hat{\boldsymbol{f}}^{(t)}$ at coordinate (k, x, y) can be computed using the nearest neighbor assignment method. The process can be written as:

$$\hat{\boldsymbol{f}}^{(t)}(k, x, y) = \arg\min_{t} \|\boldsymbol{f}^{(t)}(k, x, y) - l_t^{(t)}\|, \tag{6.16}$$

where $l_t^{(t)} \in \{l_1, l_2, \ldots, l_Q\}$ is the quantization center.

However, a major issue with in quantization is that the gradients of (6.16) are almost zeros everywhere, making gradient descent-based optimization ineffective for end-to-end communication systems. To address the zero gradient problem and approximate formulation (6.16), we introduce a differentiable soft quantization approach using continuous functions [20]:

$$\tilde{\boldsymbol{f}}^{(t)}(x, y, k) = \sum_{t=1}^{Q} \frac{exp(-\sigma \|\boldsymbol{f}^{(t)}(k, x, y) - l_t^{(t)}\|)}{\sum_{p=1}^{Q} exp(-\sigma \|\boldsymbol{f}^{(t)}(k, x, y) - l_p^{(t)}\|)} l_t^{(t)}, \tag{6.17}$$

where σ is a hyperparameter relating to the softness of the quantization. It is important to note that the quantization formulation (6.16) is used during the forward pass, while the differentiable soft quantization formulation (6.17) is used during the backward pass.

Entropy coding plays a crucial role in reducing the expected bit length by assigning short codes to frequently occurring codewords and long codes to others. In our approach, we employ Huffman coding as the entropy coding method. The codeword probability is predicted by building and maintaining a frequency table.

Consequently, $\hat{\boldsymbol{f}}^{(t)}$ is transformed into a variable-length binary code $\boldsymbol{r}^{(t)}$ of length $\ell(\boldsymbol{r}^{(t)})$ as follows:

$$\boldsymbol{r}^{(t)} := Huffman(\hat{\boldsymbol{f}}^{(t)}) \in \{0, 1\}^{\ell(\boldsymbol{r}^{(t)})}. \tag{6.18}$$

Furthermore, the semantic label map $\boldsymbol{M}_s$ is losslessly coded in vector graphic format, and the binary code is denoted as $\boldsymbol{r}_s$. The coded bitstream of the proposed RL-ASC consists of two parts: (i) the binary codes $\boldsymbol{r}_s$ with a fixed length $\ell(\boldsymbol{r}_{\boldsymbol{M}_s})$; (ii) the binary codes $\boldsymbol{r}^{(t)}$ with variable length $\ell(\boldsymbol{r}_m)$ that is adjusted by $a^{(t)}$. Therefore, the rate of the reconstructed image $\hat{I}^{(t)}$ at step m is defined as the total bits of all the binary codes $\boldsymbol{r}^{(t)}$ and $\boldsymbol{r}_s$, given by:

$$\psi(\hat{I}^{(t)}) = \ell(\boldsymbol{r}_{\boldsymbol{M}_s}) + \sum_{t=1}^{T} (\ell(\boldsymbol{r}^{(t)})). \tag{6.19}$$

Finally, the binary codes $\boldsymbol{r}^{(t)}$ and $\boldsymbol{r}_s$ are transmitted through the wireless channel and received by the decoder. Unless notified, we assume the wireless channel is lossless since the chapter mainly focuses on image semantic encoding and semantic decoding. Consequently, the receiver can achieve perfect $\hat{\boldsymbol{f}}^{(t)}$ and $\boldsymbol{M}_s$, enabling the appropriate preservation and transmission of essential semantics, thanks to the outstanding semantic extraction and bit allocation performance.

6.1.2.2 Generative Semantic Decoder

In this section, we introduce the generative semantic decoder, which decodes the received bitstreams into semantic concepts and combines them to reconstruct the image. While some learning-based codecs rely on convolutional neural network (CNNs) [6, 25, 31, 37], and others on recurrent neural network (RNNs) [27, 50], recent learning-based codecs [13, 47] have shown competitive or even superior performance compared to the classical codecs. For instance, the work in [13] proposed the use of discretized Gaussian mixture likelihoods to parameterize the distributions of latent codes, leading to a more accurate and flexible entropy model. Similarly, the work in [47] introduced a versatile deep image compression network based on spatial feature transform, enabling task-aware image compression for various tasks, such as classification and at variable rates. However, these works focus on optimizing the rate-distortion (RD) performance and do not meet the requirements of ISC.

Unlike conventional methods that primarily aim to reduce pixel errors, the proposed semantic decoder needs to be tolerant of pixel errors to some extent. Generative adversarial networks (GANs) [21] has shown impressive capabilities in synthesizing high-resolution images. Instead of precisely matching the ground truth image, GANs aim to approximate the underlying distributions of the real images.

As a result, the generated image maintains the essential meaning of the ground truth image while potentially differing in pixel values. This aligns well with the spirit of the ISC, making GANs a suitable choice for the semantic decoder in our proposed method.

Adversarial training of the GANs has shown advantages in reducing compression artifacts in deep compression systems [2, 6, 44, 48]. Motivated by existing methods, we adopt conditional GANs in the semantic decoder to translate the semantic concepts into the reconstructed image. The reconstructed image should possess naturalness and visual similarity to the input image. Naturalness is measured by the adversarial loss, which quantifies the divergence in probability distribution between real and reconstructed images, while visual similarity is captured by the aforementioned perceptual loss, which measures the distance in feature space [10].

Local Generator $\mathcal{G}_l$ The quantity and spatial occupation of different categories are imbalanced in the training dataset. It is extremely difficult to generate small object classes and texture details since the dataset is dominated by frequently occurring or large object classes. To avoid the interference from other classes, we propose a novel local image generator $\mathcal{G}_l$ to separately reconstruct the class-specific objects $\boldsymbol{y}^{(t)} \in \mathbb{R}^{H \times W}$ from the quantized class-specific feature maps $\hat{\boldsymbol{f}}^{(t)}$ as:

$$\boldsymbol{y}^{(t)} = \mathcal{G}_l(\hat{\boldsymbol{f}}^{(t)}). \tag{6.20}$$

The locally reconstructed image $\hat{\boldsymbol{x}}_l \in \mathbb{R}^{H \times W}$ is obtained by element-wise addition of all the individual $\boldsymbol{y}^{(t)}$'s as:

$$\hat{\boldsymbol{x}}_l = \boldsymbol{y}^{(1)} \oplus \boldsymbol{y}^{(2)} \oplus \cdots \oplus \boldsymbol{y}^{(M)}. \tag{6.21}$$

We feed M feature maps $\hat{\boldsymbol{f}}^{(t)}$ into a convolutional layer and five consecutive residual blocks. The outputs are then upscaled using two consecutive deconvolutional layers. Subsequently, the upscaled feature maps of M semantic concepts are individually processed by M corresponding deconvolutional layers to produce $\boldsymbol{y}^{(t)}$. Each deconvolutional layer has its own set of network parameters and effectively preserves class-specific features with rich details.

Global Generator $\mathcal{G}_g$ To capture the global structure information and the spatial layout, we employ global generation. Inspired by GauGAN [40], our well-designed $\mathcal{G}_g$ utilizes spatially adaptive normalization (SPADE) to incorporate spatial layout information from $\boldsymbol{M}_s$ into the feature maps $\hat{\boldsymbol{f}}$. The SPADE modulates the layer activation according to $\boldsymbol{M}_s$ to guide the reconstruction of different categories. The globally reconstructed image $\hat{\boldsymbol{x}}_g$ is obtained as follows:

$$\hat{\boldsymbol{x}}_g = \mathcal{G}_g(\hat{\boldsymbol{f}}, \boldsymbol{M}_s). \tag{6.22}$$

Particularly, we feed $\hat{\boldsymbol{f}}$ into three residual blocks with SPADE (SPADE-R). The output is upsampled three times to yield the global reconstruction $\hat{\boldsymbol{x}}_g$.

Attention Module In order to better combine the outputs of $\mathcal{G}_l$ and $\mathcal{G}_g$, we further propose an attention module that learns the local weight matrix $\boldsymbol{W}_l \in \mathbb{R}^{H\times W}$ and the global weight matrix $\boldsymbol{W}_g \in \mathbb{R}^{H\times W}$. The feature map $\hat{\boldsymbol{f}}$ is fed into the attention module, which consists of three consecutive deconvolutional layers and a convolutional layer. The output of the convolutional layer has two channels representing the two weight matrices and is further normalized by a softmax layer along the channel dimension. Thus, we obtain $\boldsymbol{W}_g$ and $\boldsymbol{W}_l$ whose values range between 0 and 1 and satisfy the condition $\boldsymbol{W}_l(x, y)+\boldsymbol{W}_g(x, y) = 1$ at each location (x, y). The final output image $\hat{\boldsymbol{x}}$ is obtained by element-wise summation of the weighted local reconstruction $\hat{\boldsymbol{x}}_l$ and the global reconstruction $\hat{\boldsymbol{x}}_g$:

$$\hat{\boldsymbol{x}} = (\boldsymbol{W}_l \odot \hat{\boldsymbol{x}}_l) \oplus (\boldsymbol{W}_g \odot \hat{\boldsymbol{x}}_g). \tag{6.23}$$

We employ a multi-scale patch-discriminator $\mathcal{D}$ to discriminate between the real image I and the final output image $\hat{\boldsymbol{x}}$. The generators $\mathcal{G}_l$ and $\mathcal{G}_g$ are trained alternately with the discriminator $\mathcal{D}$. We denote $\mathcal{D}(I, \boldsymbol{M}_s)$ or $\mathcal{D}(\hat{\boldsymbol{x}}, \boldsymbol{M}_s)$ as the predicted probability that the image pair $(I, \boldsymbol{M}_s)$ or $(\hat{\boldsymbol{x}}, \boldsymbol{M}_s)$ is from real samples. The discriminator $\mathcal{D}$ learns to distinguish real image pairs from fake ones by maximizing $\mathcal{D}(I, \boldsymbol{M}_s)$ and minimizing $\mathcal{D}(\hat{\boldsymbol{x}}, \boldsymbol{M}_s)$. On the other hand, the generators $\mathcal{G}_l$ and $\mathcal{G}_g$ learn to fool $\mathcal{D}$ by minimizing $(-\mathcal{D}(\hat{\boldsymbol{x}}, \boldsymbol{M}_s))$. We adopt the hinge loss for GANs, leading to the following objective functions:

$$\begin{aligned} &\min_{\mathcal{D}} L_{\mathcal{D}} = \mathbb{E}[\max(0, 1 - \mathcal{D}(I, \boldsymbol{M}_s))] + \mathbb{E}[\max(0, 1 + \mathcal{D}(\hat{\boldsymbol{x}}, \boldsymbol{M}_s))], \\ &\min_{\mathcal{G}_l, \mathcal{G}_g} L_{\mathcal{G}} = -\mathbb{E}[\mathcal{D}(\hat{\boldsymbol{x}}, \boldsymbol{M}_s)]. \end{aligned} \tag{6.24}$$

The hinge loss $L_{\mathcal{D}}$ is used for "maximum-margin" classification, penalizing the positive samples with $\mathcal{D}(I, \boldsymbol{M}_s) < 1$ and the negative samples with $\mathcal{D}(\hat{\boldsymbol{x}}, \boldsymbol{M}_s) > -1$. Only the samples that are not properly classified will impact the gradient update of $\mathcal{G}_l$, $\mathcal{G}_g$, and $\mathcal{D}$. The formulation (6.24) forces the probability distribution of $\hat{\boldsymbol{x}}$ to approximate that of the training image set. If the feature extraction network W_E is unable to preserve exact details in I, then $\mathcal{G}^l$ and $\mathcal{G}^g$ can synthesize details to match the natural image distribution, avoiding blocky and blurry effects. Equilibrium is achieved when $\mathcal{D}$ classifies the reconstructed image $\hat{\boldsymbol{x}}$ as real.

The final reconstruction $\hat{\boldsymbol{x}}$ should not only be consistent with the natural image distribution, as desired by traditional GANs, but also fulfill the requirements of semantic communications by recovering the semantic concepts of a specific input image I. To guide the semantic reconstruction process and ensure that $\hat{\boldsymbol{x}}$ approximates the feature space of I, we employ the perceptual loss (6.6).

The proposed semantic encoder and semantic decoder are jointly trained, and the objective function can be formulated as the weighted combination of adversarial loss $L_{\mathcal{D}}$, perceptual loss L_P, and feature classification loss L_C as:

$$\min_{W_E, \mathcal{G}, \mathcal{D}} L_{\mathcal{D}} + \lambda_1 L_P + \lambda_2 L_C, \tag{6.25}$$

where λ_1 and λ_2 are the weighting parameters. Due to the well-preserved semantic information, the proposed generative semantic decoder outperforms leading image generation models [1, 2, 40]. Notably, the work in [1] poorly preserves appearance features, resulting in a reconstructed image that deviates significantly from the source image. The work [40] can only generate general images from the distribution of the training dataset without controlling the appearance of specific objects.

6.1.3 Experimental Results

6.1.3.1 Simulation Setup

Datasets We train and evaluate the RL-ASC model based on the scene parsing and instance segmentation of the Cityscapes dataset [17]. Cityscapes focuses on the semantic understanding of urban street scenes. It is collected from streetscapes in 50 different German cities and consists of 30 classes of objects. There are 2,975 images in the training set and 500 images in the validation set, with each image annotated with fine semantic labels. We downscale the images and semantic label maps to 256×512 and conduct testing on the validation set. The classes or the semantic concepts in the Cityscapes dataset are listed as Table 6.1.

Baselines We compare the proposed RL-ASC with the engineered codecs BPG [11], JPEG2000 [42], JPEG [52], and the deep learning–based codecs deep semantic segmentation-based layered image compression (DSSLIC) [2] and high-fidelity generative image compression (HiFIC) [36]. Moreover, we fine-tune the DSSLIC with the semantic loss in the semantic segmentation task for a fair comparison, and the fine-tuned model is denoted as DSSLIC-fine-tuned. The bit rate of the engineered codecs [11, 42, 52] is controlled by the quantization parameters (QP), where a larger QP corresponds to a higher compression ratio. We use the official versions of these codecs and evaluate their performance in the bit rate range of 0–0.5 bpp. BPG [11] is currently the state-of-the-art engineered image compression

Table 6.1 The classes in the Cityscapes dataset

Group	Classes
Flat	road · sidewalk · parking · rail track
Human	person · rider
Vehicle	car · truck · bus · on rails · motorcycle · bicycle · caravan · trailer*+
Construction	building · wall · fence · guard rail · bridge · tunnel
Object	pole ·pole group · traffic sign · traffic light
Nature	vegetation · terrain
Sky	sky
Void	ground · dynamic · static

codec in terms of PSNR. DSSLIC [2] is a layered image compression method, where the semantic label of the input image is encoded as the base layer of the bitstream, while the compact representation and residual are encoded as the enhancement layer. The compression ratio of DSSLIC and DSSLIC-fine-tuned is adjusted by the QP of the enhancement layer. HiFIC [36] combines GANs with learned compression to achieve high-fidelity generative lossy compression, providing visually pleasing reconstructions that are perceptually similar to the input image across a broad range of bit rates. We evaluate the performance of [36] by leveraging the pretrained models at low bi trate (0.18 bpp) and medium bit rate (0.33 bpp). It should be noted that it is challenging to tune the same bit rate for all baselines and the proposed method. Therefore, we run the baselines and compare the reconstruction and semantic performance of different methods at similar bit rates.

6.1.3.2 Evaluation Metrics

To evaluate the efficiency of the proposed RL-ASC, we assess the reconstruction performance from both semantic and perceptual perspectives. Traditional quality metrics such as PSNR and structural similarity (SSIM) are simplistic and fail to capture certain nuances of human perception. On the one hand, we evaluate the semantic loss based on downstream tasks such as object detection and semantic segmentation. We adopt mean intersection over union (mIoU) as the objective metric and also illustrate the results of semantic segmentation and object detection as the subjective evaluation. A reconstructed image that retains more semantic information will yield a higher mIoU value, with a maximum value of 1.

On the other hand, the metrics Fréchet Inception distance (FID) score [24], and Kernal inception distance (KID) [8] (lower better) are consistent with human perception and thus are employed to evaluate the distance of the reconstructed and input images in deep feature space. Specifically, KID [8] and FID [24] measure the distribution divergence of the reconstructed images compared with real samples via the inception network and are widely used to assess sample quality and diversity in the context of GANs. Moreover, the perceptual performance can also be evaluated subjectively by user studies, and a better compression approach can yield reconstructions that appear real, natural, and visually pleasing even at low bit rates.

Compression Modes We train three bit rate models for RL-ASC: the low bit rate (0.08 bpp), the medium bit rate (0.16 bpp), and the high bit rate (0.32 bpp). The bit rate is adjusted by the channel dimension n of the feature map $\boldsymbol{f}$ that is produced by the feature extraction network W_E. We set $n = 16, 32, 64$, $w = \frac{W_{\text{image}}}{8}$ and $h = \frac{H_{\text{image}}}{8}$, resulting in $\boldsymbol{f}$ dimensions of $16 \times \frac{W_{\text{image}}}{8} \times \frac{H_{\text{image}}}{8}$, $32 \times \frac{W_{\text{image}}}{8} \times \frac{H_{\text{image}}}{8}$ and $64 \times \frac{W_{\text{image}}}{8} \times \frac{H_{\text{image}}}{8}$ for the three models, respectively. The dimension of the downscaled semantic mask $\boldsymbol{M}_{s_d}^{(t)}$ is $\frac{W_{\text{image}}}{8} \times \frac{H_{\text{image}}}{8}$. We set $Q = 6$, allowing the RL agent π to choose six quantization levels for different semantic concepts, where a higher quantization level indicates lower distortion.

To evaluate the impact of the RL-based semantic bit allocation model, we conduct an ablation study by removing the RL agent π from the RL-ASC model. Such an ISC system lacks the adaptive bit allocation ability and thus is denoted as simplified RL-ASC. In simplified RL-ASC, each semantic concept is encoded with equal precision using the highest quantization level. The three bit rate models for simplified RL-ASC are low bit rate (0.11 bpp), medium bit rate (0.22 bpp), and high bit rate (0.44 bpp).

6.1.3.3 Semantic Performance

We compare the semantic performance of the proposed RL-ASC and the simplified RL-ASC with the baseline codecs at different bit rates. To validate the effectiveness of the proposed RL-ASC, we apply the method to two downstream semantic tasks: semantic segmentation and object detection.

Objective Quality We use the pretrained PSPNet [61] as the semantic segmentation model to obtain the semantic label $\hat{s}$ of the reconstructed image. The semantic fidelity is measured by the consistency between $\hat{s}$ and the ground truth $\boldsymbol{M}_s$ in terms of mIoU. The mIoU performance of different image codecs in different bit rates is shown in Fig. 6.4. The Cityscapes dataset only provides ground truth labels for semantic segmentation and lacks ground truth labels for other intelligent tasks such as object detection and image classification. Therefore, we cannot obtain the

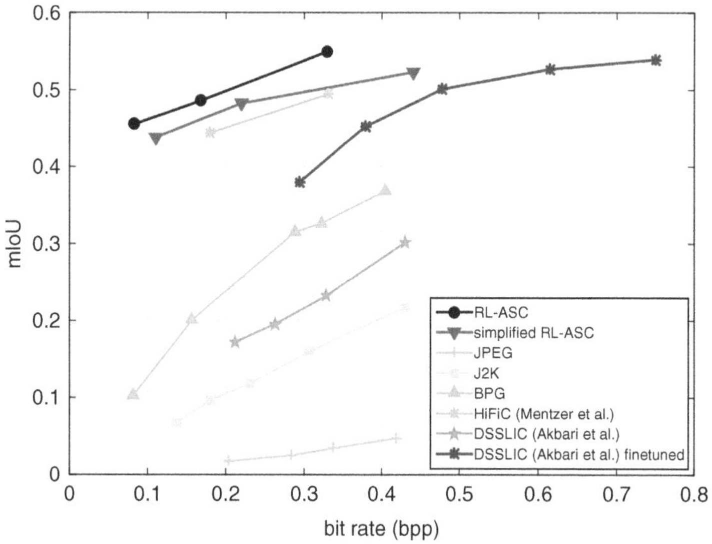

Fig. 6.4 The semantic performance in terms of mIoU of different image codecs on semantic segmentation task. A higher value means better performance

mean average precision (mAP) metric of the object detection task for objective measurement.

From Fig. 6.4, it can be observed that semantic fidelity increases with higher bit rates. The proposed RL-ASC outperforms JPEG [52], J2K [42], BPG [11], DSSLIC [2], and DSSLIC-fine-tuned by a large margin in terms of mIoU, validating the effectiveness of the proposed task-driven coding approach. The baseline HiFiC [36] approximates the performance of the proposed RL-ASC, but fails to achieve extremely low bit rates (<0.1 bpp). RL-ASC achieves higher mIoU performance compared to simplified RL-ASC. It is worth noting that DSSLIC fine-tuned with semantic loss improves the performance in terms of mIoU for intelligent tasks.

We use the Bjontegaard metric [9] to evaluate the coding efficiency of the proposed RL-ASC compared to the baselines. Inspired by Li et al. [33], we propose BD-mIoU to consider the relative differences in task-related accuracy between two codecs under equal bit rate. BD-mIoU calculates the average mIoU difference between two rate-semantic curves over an interval, and BD-rate represents the average bit rate reduction under the equivalent task-related accuracy. Table 6.2 shows that the proposed RL-ASC achieves the same mIoU with more than 60% bit rate savings on average compared to the deep learning–based methods [2, 36] and DSSLIC-fine-tuned. Under the same bit cost, the proposed RL-ASC method significantly improves the mIoU performance compared to the baselines and simplified RL-ASC.

Subjective Quality In this section, we visualize the semantic segmentation and object detection results of the decompressed images to subjectively validate the semantic fidelity. To ensure a fair comparison, the proposed RL-ASC encodes the image at a bit rate of 0.33 bpp, while the baselines encode the image at an equal or higher bit rate. We adopt the MaskRCNN [23] pretrained on the COCO dataset to detect objects in the reconstructed image.

Figure 6.5 shows the input images, ground truths, and the predicted semantic label maps on the decompressed images. It can be observed that the semantic concepts in the proposed RL-ASC are well recognized and localized in the decompressed images (a) and (g). The segmentation results are similar to the ground truths, validating that the proposed RL-ASC can preserve the overall meanings of the input image with minimal semantic information loss. The deep learning–based codecs [2] and HiFiC [36] outperform the classic engineered codecs [11, 42, 52]. The semantic labels (b), (c), (h), and (i) represent major concepts such as cars, buildings, and roads, while less important information is missing. However, JPEG [52], J2K [42], and BPG [11] fail to perform the semantic segmentation task, as

Table 6.2 BD-mIoU and BD-rate relative to the baselines for semantic segmentation tasks

Metric	DSSLIC [2]	DSSLIC-fine-tune	HiFiC [36]	BPG [11]	J2K [42]	JPEG [52]	Simplified RL-ASC
BD-mIoU	0.278	0.145	−0.004	0.227	0.343	0.452	0.0261
BD-rate	−97.713%	−60.769%	9.951%	−89.682%	−99.688%	−100%	−31.361%

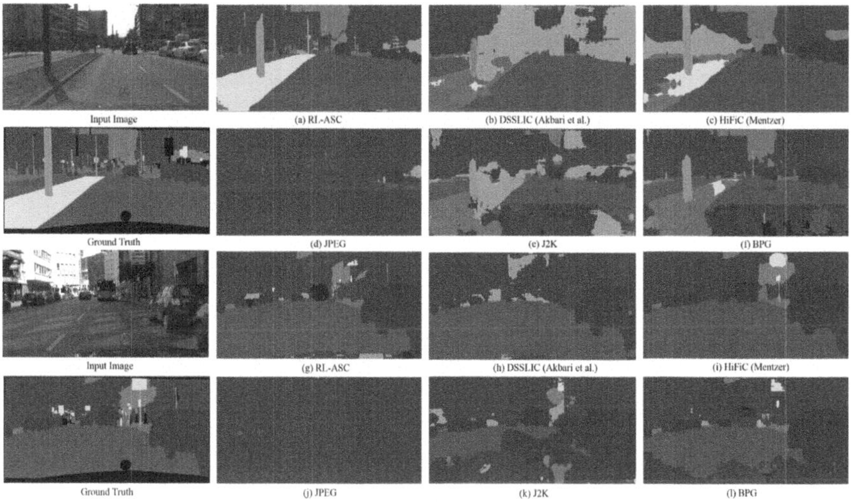

Fig. 6.5 Examples of image coding on downstream semantic segmentation. The first column contains the two randomly selected input images before coding and the corresponding ground truth labels. The rest of the columns are the semantic label maps of decompressed images of different codecs under similar bit rate (0.33 bpp)

shown in (d), (e), (f), (j), (k), and (i), and the semantic concepts are misinterpreted by the downstream task. This degradation is due to blocky, blurring, or ringing artifacts in the decompressed images of the engineered codecs at low bit rates.

The object detection performance on the decompressed images of the proposed RL-ASC and the baselines is illustrated in Fig. 6.6. It can be observed that the decompressed images (a) and (g) of the proposed RL-ASC preserve the objects comprehensively. Small objects such as traffic lights and overlapped cars can be properly detected in (a) and (g), demonstrating the remarkable semantic exchange ability of the proposed RL-ASC method. However, since the baselines aim to recover the exact pixels while ignoring the global underlying meanings, some reconstructed objects fail to be accurately detected. For instance, the traffic light is not detected in the HiFiC decompressed image (c) and the DSSLIC decompressed image (h), and larger objects such as cars and persons may even be misdetected in the classic engineered codecs [11, 42, 52].

6.1.3.4 Perceptual Performance

We compare the perceptual performance of the proposed RL-ASC and simplified RL-ASC with the baselines at different bit rates.

Objective Quality The perceptual performance of different image codecs in terms of FID [24] and KID [8] at different bit rates is illustrated in Fig. 6.7. The proposed

Fig. 6.6 Examples of image coding on downstream object detection task. Two randomly selected images are given as examples. The bounding boxes and labels of the decompressed images of different image codecs are produced by the pretrained MaskRCNN

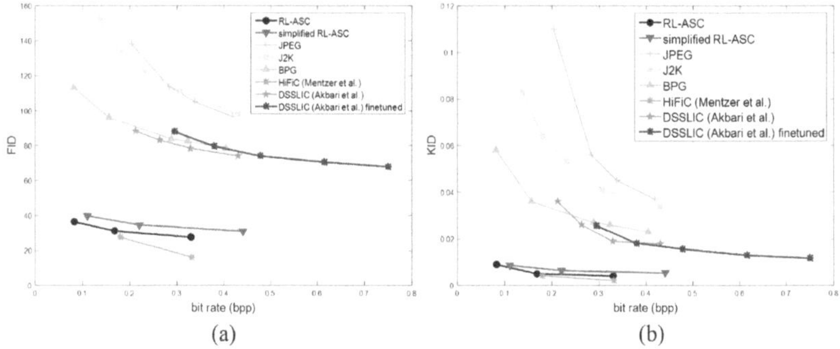

Fig. 6.7 The perceptual performance of different image codecs in terms of (**a**) FID and (**b**) KID. The lower value means better performance

RL-ASC is comparable to HiFiC [36] and outperforms the other baselines by a large margin. This can be attributed to the incorporation of GAN architecture and the adoption of adversarial loss, which enforce the reconstructed image to be natural and realistic. This result validates the effectiveness of the well-designed semantic

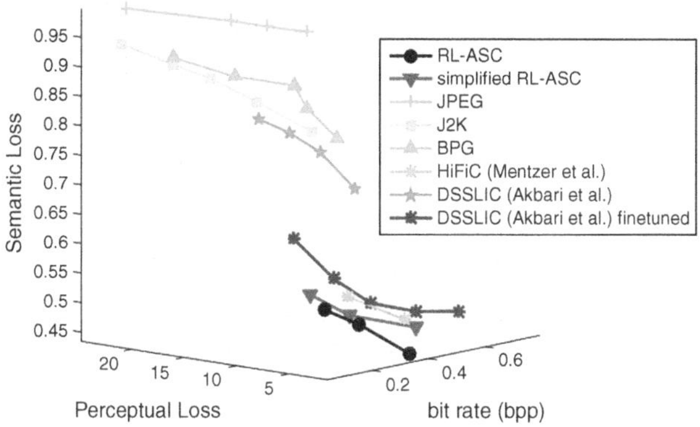

Fig. 6.8 The rate-semantic-perceptual curves of different image codecs. At a certain bit rate, lower semantic loss and lower perceptual loss mean better performance

encoder and semantic decoder. Furthermore, the RL-based semantic bit allocation model leads to a significant improvement in perceptual performance compared to the simplified RL-ASC. There is a moderate degradation in FID and KID for the fine-tuned DSSLIC compared to the original DSSLIC model, which can be attributed to the fine-tuned model's emphasis on semantic fidelity at the expense of perceptual performance.

Figure 6.8 illustrates the triple trade-off between rate, semantic, and perceptual performance of the proposed RL-ASC method and the baselines. Specifically, the semantic loss is defined as (1-mIoU), taking into account the semantic segmentation task. The perceptual loss is defined as the weighted addition of FID and KID values, where a lower value indicates better performance. As shown in Fig. 6.8, the proposed RL-ASC achieves lower semantic and perceptual loss compared to the baselines at an equal bit rate, validating the effectiveness of the proposed method in semantic exchange and visual performance. It is worth noting that the deep learning–based methods RL-ASC, [2, 36] outperform the classic engineered codecs, indicating that deep features contribute to better image understanding. Moreover, DSSLIC fine-tuned with semantic loss achieves significant improvement in this triplet loss compared to the original DSSLIC model.

Subjective Quality Figure 6.9 shows the reconstructed images from different image codecs, as well as the original randomly selected input image. To ensure a fair comparison, RL-ASC encodes the image at a bit rate of 0.16 bpp, while the baselines encode the image at an equal or higher bit rate. It can be observed that the decompressed image (b) of the proposed RL-ASC is almost indistinguishable from the input image (a) even at such a low bit rate. Compared to the simplified RL-ASC (c), the bits in (b) are concentrated on salient semantic concepts, resulting in visually pleasing reconstruction. In contrast, the baselines suffer from blur, ringing, and blocky artifacts at low bit rates, as shown in (d), (f), (g), and (h).

Fig. 6.9 A randomly selected input image and the reconstructed images of different image codecs at similar bit rate (0.16 bpp)

6.2 Rebuffering Optimization for DASH via Pricing and EEG-Based QoE Modeling

Pricing is an effective mechanism for network resource allocation that can achieve a desirable balance between efficiency and fairness. Among various performance indices of video transmission, rebuffering is an important factor that influences user quality of experience (QoE). Therefore, we propose a price-based bandwidth allocation scheme for dynamic adaptive streaming over hypertext transfer protocol (DASH) systems to mitigate the effect of rebuffering on QoE. The utility model of the proposed scheme considers the relationship between allocated bandwidth and rebuffering length, as well as the impact of rebuffering length on QoE. Specifically, we conduct electroencephalography (EEG) experiments and calibrate the distribution of the subjects' time limits at which rebuffering triggers negative emotions. Based on this model, the DASH server periodically collects buffer state information from all DASH clients to adjust its bandwidth allocation. Assuming the server protects itself from congestion by pricing the clients' requested bandwidth, we formulate a Stackelberg game to study the joint utility maximization of the server's revenue and the clients' utility.

In the context of DASH, the video content is encoded at multiple bit rates and segmented into chunks of identical playout duration. The client determines the bit rate of the next chunk to download based on a rate decision algorithm (RDA) and the collected channel and buffer information. However, parameters visible to the client, such as rebuffering length, are not directly related to the allocated resources, and sophisticated models are required to evaluate the relationships between these visible parameters and communication resources.

Most studies on QoE optimization in DASH systems, including those considering rebuffering, focus on sophisticated RDAs on the client side [28, 43, 57, 58]. For example, in [28], the rebuffering percentage during the playback of multiple chunks is estimated and kept below a threshold to optimize QoE. The only study that models rebuffering length is [57], which considers the total lengths of rebuffering events in the QoE function and optimizes the sum of QoE over a period for a single-user DASH system. Furthermore, calibrating these visible parameters in terms of user QoE is a complex issue. Currently, the most common approach [12, 26] for QoE assessment is a judgment experiment in which subjects rate the quality of test videos. However, such ratings may be influenced by individual biases and strategies [45].

Recently, psychophysiological methods have been considered more promising than psychophysical judgment experiments for measuring QoE. In particular, EEG, a noninvasive technique for measuring voltage changes on the scalp with high temporal precision, has been adopted in various video QoE assessment studies [4, 29, 38, 45]. EEG data can be analyzed in two main ways: event-related potential (ERP) and spectral analysis [5]. ERP is a time-locked reaction in terms of voltage change and can be correlated with the level of visual distortion [45] or blockings [4]; However, due to its high signal-to-noise requirement, the video stimuli in those experiments must be short and have little semantic content, which does not align with realistic settings. Alternatively, spectral analysis is less sensitive to noise, and the power levels of certain frequency bands have been found to correlate well with human cognitive and emotional processes.

6.2.1 System Model and Rebuffering Length Calculation

As shown in Fig. 6.10, we consider a system consisting of a DASH server and N_U DASH users, each of which requests a video flow from the server. We assume that the main bottleneck of transmission is the export bandwidth of the DASH server.

The videos are encoded into N_Q profiles of N_Q bit rates. The bit rates are $\varphi_1 \leq \varphi_2 \leq \ldots \leq \varphi_{N_Q}$, each of which represents a quality level. Each profile is segmented into chunks of playback duration T_C. After completely receiving a chunk, the user

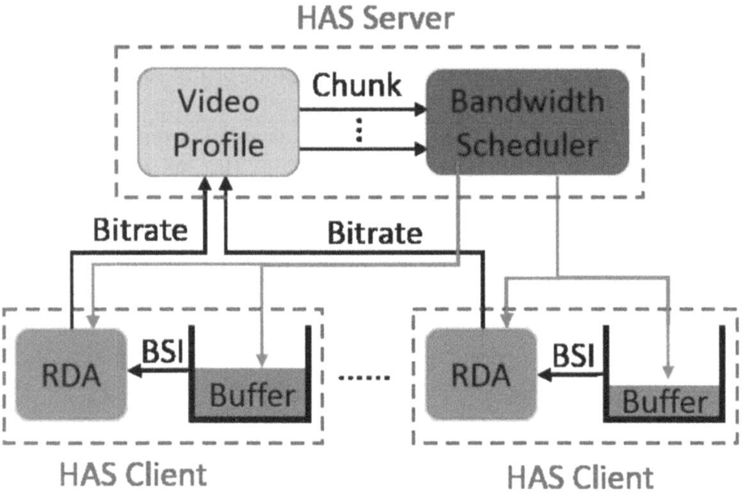

Fig. 6.10 System model. We consider a system consisting of a DASH server and N_U DASH users, each of which requests a video flow from the server. We assume that the main bottleneck of transmission is the export bandwidth of the DASH server. After completely receiving a chunk, the user chooses a bit rate for the next chunk to download according to a certain RDA and sends its request to the server with its current state

chooses a bit rate for the next chunk to download based on a certain RDA and sends the request to the server along with its current state.

The server periodically adjusts the bandwidth allocation scheme periodically (at a time scale of T_S) based on the collected buffer state information (BSI) and utility model. Since the length of each chunk in most DASH literature is larger than 2 s [14, 28], the time period for updating bandwidth allocation can be less than 1 s. Therefore, we assume that $T_C > 2T_S$. When $t = \tau T_S$, the server collects the BSI of each user i is collected by the server to update its bandwidth allocation scheme for the next slot $\tau + 1$. The BSI includes the current rebuffering length $RBU[\tau T_S]$, the remaining buffer length $BU[\tau T_S]$, the bit rate of the chunk currently downloading chunk $\varphi_i[\tau T_S]$, and the already downloaded length of the chunk currently downloading chunk $DU[\tau T_S]$.

The DASH client of each user continues downloading regardless of the buffer state; however, playback is suspended when the buffer is exhausted (i.e., $B_i(t) = 0$), and resumes after the current chunk is downloaded (i.e., $D_i(t) = T_C$). The bandwidth allocation of each user is fixed in duration $[\tau T_S, (\tau+1)T_S]$. Therefore, by using the collected BSI, we can derive the relationship between rebuffering length and allocated bandwidth for this slot, i.e., $RB_i[(\tau + 1)T_S]$ and $R_i[(\tau + 1)T_S]$ can be derived as follows:

(a) $RB_i[\tau T_S] = 0, B_i[\tau T_S] \geq T_S$. In this solution, no rebuffering occurs at the beginning, and the buffer length is greater than T_S. Therefore, no rebuffering occurs because the video of the next playback length T_S is already downloaded.

(b) $RB_i[\tau T_S] = 0, B_i[\tau T_S] \leq T_S, T_S + D_i[\tau T_S] - B_i[\tau T_S] \leq T_C$. The buffer length is less than T_S, and the playback video of the next T_S length has the same quality as the currently downloading video. To avoid rebuffering, a video of length $T_S - B_i[\tau T_S]$ must be downloaded within the next slot.

6.2.2 Calibration of the Effect of Video Rebuffering on QoE Using EEG

Humans have a time limit for reacting to different stimuli [49, 51]. Therefore, we aim to determine a time limit at which subjects begin to experience negative emotions due to a video rebuffering event. As long as the rebuffering ends before that limit, it does not degrade the user experience. Based on these findings, we present a QoE model for rebuffering length that takes into account the distribution of calculated time limits.

6.2.2.1 Methods

Participants A total of eight healthy volunteers (four female, four male) aged between 18 and 28 years old participate in the experiment. All participants have normal or corrected-to-normal vision and report no history of neurological diseases.

Stimuli, task, and procedure: Animal documentaries are used to create the stimuli as they are unlikely to cause emotional changes, ensuring a high signal-to-noise ratio. A total of 170 nonoverlapping 30 s video clips are extracted from three documentaries for further processing. The video clips are randomly divided into five groups: one control group without rebuffering and four test groups with rebuffering events of 1, 3, 5, and 7 seconds. Rebuffering begins randomly according to a uniform distribution at the interval of 10 s–20 s to prevent participants from predicting the time point of the start or the end of rebuffering. The processed video clips are trimmed to a length of 30 s by discarding the frames beyond that point.

The scalp gel is prepared to ensure a resistance below 10kΩ between each electrode and the scalp. After conducting vision acuity and Ishihara color blindness tests, six blocks of experimentation are performed, consisting of 25, 30, 30, 28, 28, and 29 trials. In each trial, participants are first shown attention words, followed by the display of a 30 s video clip. For 55 out of 170 trials, a question is presented, and participants have to press a Y/N button to indicate their answer. For the remaining 115 trials, participants simply press a Y button when they are ready for the next trial, which start immediately after the button press. Each block lasts approximately 20 min and is followed by a 5 min break (Table 6.3). The procedure of one section of the experiment is shown in Fig. 6.11.

EEG data acquisition Brain activity is recorded using a 64-channel EEG amplifier (BrainAmps, Brain Products, Germany) and a Brain Vision Recorder (Brain Prod-

Table 6.3 Setup of EEG experiments

Type	Term	Value
Stimuli	Video source	Life story
	Resolution	720×404 pixels
	Frame rate	25 fps
Test environment	Resolution of monitor	1920×1080
	Refresh rate	144 Hz
	Display window	16×9 cm
	Visual angle	9 degree

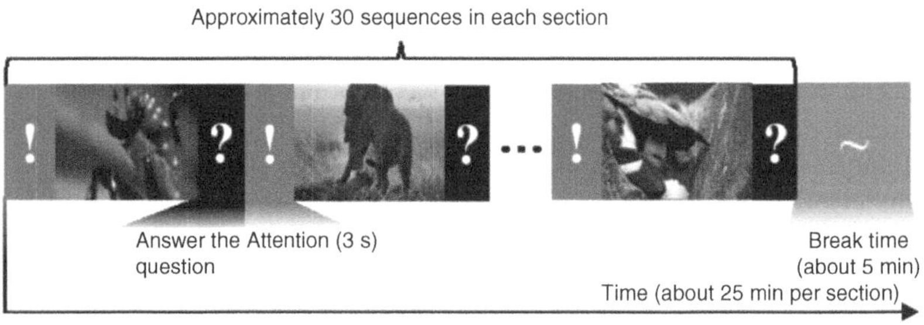

Fig. 6.11 Procedure of one section of the experiment

ucts, Germany) with a sampling rate of 500 Hz and a bandpass filter of 0.531–70 Hz. The EEG cap consists of 64 actiCAP active electrodes (Brain Products, Germany) positioned according to the 10–20 system. The ground electrode is placed between electrodes Fp1 and Fp2, and electrode Cz serves as the reference.

EEG preprocessing All analyses are performed using MATLAB 2014b (MathWorks, Natick, Massachusetts). EEG data are first re-referenced to the average of the left and right earlobes, followed by the application of a 0.1 Hz high-pass filter (FIR). Continuous EEG data are epoched from −4 s to 9 s from the onset of rebuffering (control group trials are epoched at a random time point from 10 s to 20 s). Next, manual trial rejections are performed by excluding trials with eye blinks at the onset of a rebuffering event or drifts during the process, resulting in an average rejection rate of 20% per participant. Visual examination of the independent component analysis (ICA) components is conducted to identify and remove components capturing eye blinks.

Time-frequency analysis To examine the spectral power changes caused by rebuffering, the time series data are decomposed into time-frequency representations [16]. The time series is convolved with a set of Morlet wavelets of frequencies ranging from 1 to 30 Hz in 60 linearly spaced steps. The complex wavelets are created by multiplying complex sine waves ($e^{i2\ ft}$) where i is the complex operator, f is frequency, and t is time) by a Gaussian shape function ($e^{t^2/2\sigma_t^2}$), where σ_t is the width of the Gaussian shape). The width of the Gaussian shape function is set as $_t = c(f)/(2\ f)$, where $c(f)$ represents the number of cycles of each wavelet.

$c(f)$ is linearly spaced between 2 and 9 to achieve a balance between temporal and frequency precision. The power at each time-frequency points $P(t, f)$ is defined as the squared magnitude of the complex signal resulting from the convolution. The power is then normalized in decibels as follows:

$$dBP(t, f) = 10 \log_{10} P(t, f)/Base(f), \tag{6.26}$$

where $Base(f)$ represents the average power in the period from -4000 ms to -2000 ms from the onset of rebuffering for each channel, frequency, and trial.

Selection of frequency bands and channels Our goal is to identify frequency and channel components that exhibit significant changes during rebuffering periods In addition, the changes we find should be common among most subjects. To achieve this, we perform statistical tests comparing rebuffering and playback signals to find individual components with significant differences and identify the most common components across all subjects.

Playback signals are generated using the time-frequency maps of the control group and the time-frequency maps of -4 s to -1 s from the onset of rebuffering in the test groups. Rebuffering signals are created using maps from the rebuffering time courses of all test groups. The signals are averaged over each selected time course to obtain power spectral curves. For each subject, the number of playback curves should not exceed 170 (34 from the control group and 34 multiplied by 4 from the test groups), and the number of rebuffering curves should not exceed 136 (34 multiplied by 4 from the test groups).

In the first step of selection, the curves are averaged over channels to focus on the power distribution of frequency. A two-tailed t-test [30] is performed on all frequency points between the rebuffering and playback signals to obtain corresponding p values. Subsequently, a one-tailed Wilcoxon signed-rank test [30] is conducted on the p values of all frequencies over all the subjects to identify frequency points with medians of the p values significantly less than 0.05. Frequencies showing a significance level of $p < 0.05$ in the Wilcoxon signed-rank test [30] are selected as they consistently exhibit significant differences between rebuffering and playback signals. Post hoc analysis is conducted to confirm the appropriateness of the selection by comparing the distribution of individual trial-averaged mean power within the selected frequency range for rebuffering and playback signals.

In the second step of selection, channels are determined in a similar manner to the first step. The samples are averaged over the selected frequency range, and a two-tailed t-test is performed on all channels. The number of subjects showing a significance level of $p < 0.0005$ is counted for each channel. Channels with a number of significant subjects greater than or equal to 6 are selected. Again, post hoc analysis is conducted by comparing the distribution of individual trial-averaged mean power for the selected channels and frequency bands.

Calculation of time limits After determining the frequency bands and channels, individual time limits are calculated using permutation testing. All time-frequency maps are averaged over the selected channels and frequencies to obtain the temporal fluctuation of spectral power. Permutation testing [30] is then conducted on the time series of control group trials and test group trials to determine the significance of the difference at each time point. For each permutation, the trial series are randomly relabeled as either test or control group, and 1000 permuted difference signals of mean power are created. At each time point, the true observed difference between test and control group trials is subtracted from the mean permuted difference and divided by the standard deviation of the permuted differences to obtain a z value. A z value corresponding to $p < 0.05$ is chosen as the threshold of significance. The first time point after the onset of rebuffering that reaches this significance level is defined as the individual reaction time limit.

6.2.2.2 Results

Finding 1: EEG oscillations from 15–24 Hz increase during rebuffering events.

Analysis Some subjects exhibit a significant increase in mean power from high alpha to beta (Fig. 6.12a, 10–25 Hz, two-tailed t-test, $p < 0.05$) during rebuffering, while others also show an increase in lower frequencies (Fig. 6.12b, 3–4 Hz and 16–24 Hz, two-tailed t-test, $p < 0.05$). Although the frequency ranges differ slightly, they overlap in some regions. The significance levels of differences among all individuals are tested (Fig. 6.12c above, two-tailed t-test), and 15–24 Hz consistently shows a high level of significance at $p < 0.05$ (Fig. 6.12c below, one-tailed Wilcoxon signed-rank test, $p < 0.05$). Moreover, post hoc analysis reveal that the distribution of mean power in this frequency band significantly differs between the rebuffering and playback signals for all subjects (Fig. 6.12d, two-tailed paired t-test, $p < 104$). These results help us choose most of the beta-band (15–24 Hz) as the frequency range, in which a reaction to rebuffering is induced.

Previous studies have associated beta-band oscillations with attentional processes and emotional evocations [3, 7, 19, 59]. Since rebuffering affects the video service experience of subjects who are unprepared for any tasks or unaware of the existence of rebuffering, we believe that the beta-band oscillations observed in our experiments are related to emotional processes.

Finding 2: EEG oscillations around the parietal lobe increase during rebuffering events.

Analysis We further investigate the spatial distribution of this difference by examining the brain regions that consistently exhibit a significant difference in mean power within the selected frequency band (15–24 Hz) between the two types of signals. For some subjects, the reactions are concentrated in the parietal lobe (five subjects, e.g., subject #3, Fig. 6.13a), while other subjects also show an increase in power in the frontal lobe and other brain regions (three subjects, e.g., subject #5, Fig. 6.13b). The significance level is thresholded (two-tailed t-test, $p < 0.005$), and the number of subjects showing a significant difference on each channel is counted.

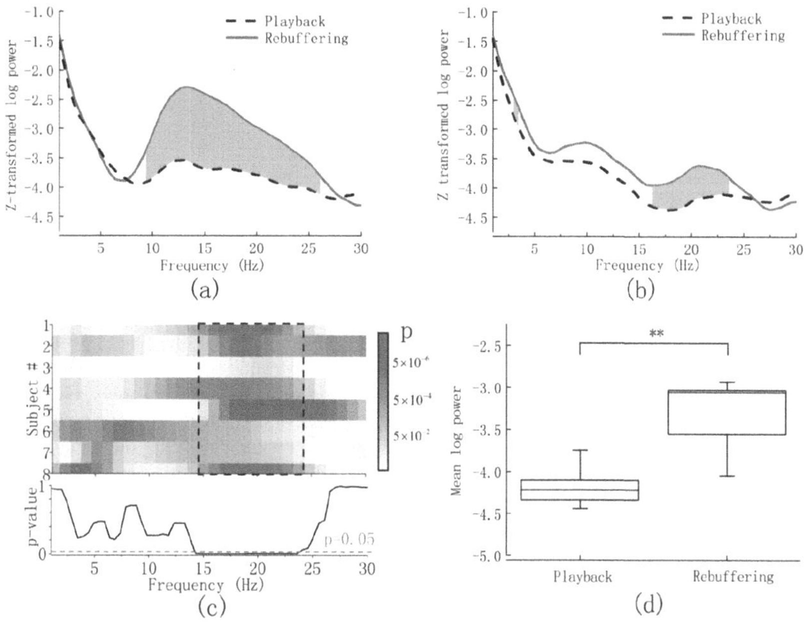

Fig. 6.12 Frequency selection. (**a**, **b**) Mean log power spectra of two subjects (#1 and #8). In each graph, the lines indicate the results of two conditions (black for playback and red for rebuffering). The shaded region indicates the frequency bands of the significant difference between the two conditions (two-tailed t-test, $p < 0.05$). (**c**) (Above) Significance levels of the difference (two-tailed t-test) between two conditions over 8 subjects and 60 frequencies (from 1 to 30 Hz, linearly spaced) and (below) the statistical results of whether 8 t-test p values at each frequency point are less than 0.05 (one-tailed Wilcoxon signed-rank test); the dashed gray line in the bottom figure indicates the threshold $p = 0.05$, whereas the dashed box in the top figure indicates the frequency band we chose according to the p values shown in the bottom curve. (**d**) Box chart of the mean power of the selected frequency band (15–24 Hz) between the playback and rebuffering signals. ** indicates that the mean power of the selected frequency band of the rebuffering signals is significantly different from the ones of the playback signals of eight subjects (two-tailed paired t-test, $t(7) = -7.607$, $p < 110^{-4}$)

In 11 channels around the parietal lobe, no fewer than seven subjects exhibit significant differences. Additionally, post hoc analysis reveal that the distributions of mean power in these channels significantly differ between the two types of signals (Fig. 6.13c, two-tailed paired t-test, $p < 107$). In conclusion, the consistent reactions observed in these 11 channels around the parietal lobe indicate that they are reasonable spatial features induced by video rebuffering.

Some studies have demonstrated the association between negative emotions and the parietal cortex [39, 46]. For instance, Davidson et al. [46] found that the presentation of an angry face induced significantly more asymmetrical parietal beta-band power than the presentation of a neutral face. This suggests that angry feelings might cause parietal beta-band oscillations. In addition, Wendy et al. [39] decomposed negative emotions into subtypes and analyzed their regional sources by

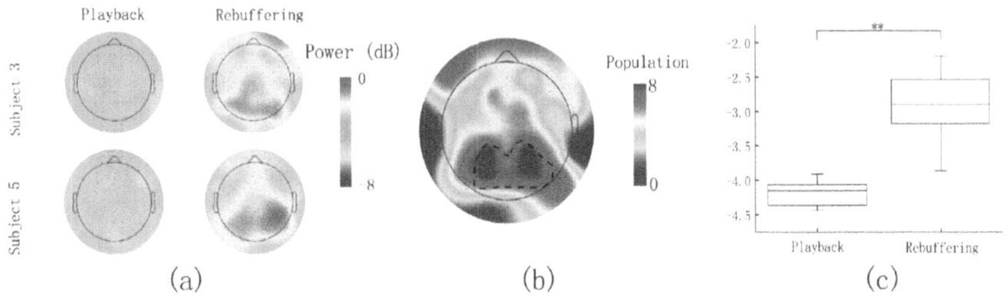

Fig. 6.13 Channel selection. (**a**) Topographical maps of the mean log power in the frequency range from 15 to 24 Hz of the playback and rebuffering signals of two subjects (#3 and #5). (**b**) Topographical maps of the number of subjects who show significant differences of mean log power of the frequency range from 15 to 24 Hz (two-tailed t-test, $p < 0.0005$) at each channel. The dashed circle surrounds the channels with a population of n $\geqslant$ 7. The channels selected are Pz, 1–6, 8; POz, 3–4, 7–8. (**c**) Box chart of the mean power of the selected frequency band and channels between playback signals and rebuffering signals. ** indicates that the values of the rebuffering signals are significantly greater than those of the playback signals (two-tailed paired t-test, $t(7) = -8.9$, $p < 110^{-4}$)

summarizing previous studies. The parietal lobe, according to this study, is strongly associated with anxiety. Furthermore, spatial working memory was found to be more vulnerable to negative emotion than verbal working memory. This means that negative emotion and spatial working memory may compete for attentional resources [32]. Because the parietal cortex is known to control spatial attention, it is possible that negative emotion originates from the parietal lobe [32].The above studies justify the conclusion that the parietal reactions observed in our experiment reflect negative emotion.

Finding 3: EEG oscillations from 15–24 Hz around the parietal lobe increase from 190 to 850 ms after the onset of a rebuffering event

Analysis After selecting the frequency band and channels, time-frequency power maps of the selected range are averaged to focus on the dynamics of brain activity intensity around the onset of a rebuffering event and determine the starting time of oscillatory activities. Individual time limits are derived by identifying the earliest time point at which the rebuffering and playback trials exhibit significant differences in signals (Fig. 6.14a, two-tailed permutation test, $p < 0.05$). Topographical maps around the time point are also inspected to avoid mistakes caused by random noise (Fig. 6.14b). For all subjects, reactions initiate from parts of the parietal lobe, which confirms our channel selection. All subjects exhibit reactions originating from parts of the parietal lobe, confirming the selection of channels. While all subjects show a continuous increase in power after the onset of a rebuffering event, the time at which the increase become significant vary from 190 ms to 850 ms (median: 470 ms, Fig. 6.14c).

Decades ago, the speed of human visual perception was determined. The earlier, weaker differential activity emerged approximately 75 ms after the presentation

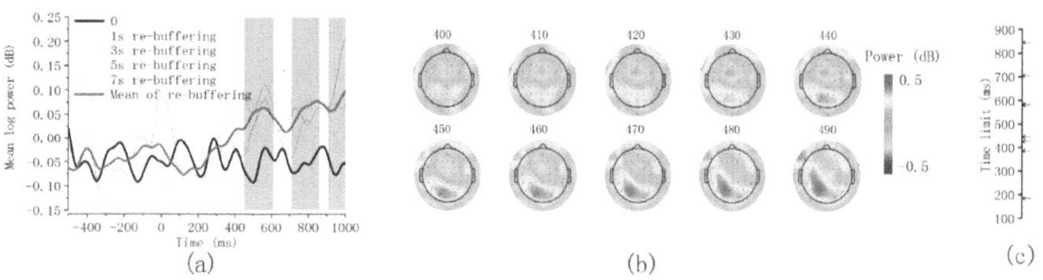

Fig. 6.14 Calculation of time limits of reaction onsets. (**a**) Mean power signal averaged over the selected frequency range and channels from −500 ms to 1000 ms of subjects #4. The thick red and black lines indicate the averaged power of rebuffering signals and playback signals respectively, whereas the four dim lines indicate the averaged power of 1-/3-/5-/7-s rebuffering events. The shaded region indicates the time windows of significant difference between the two conditions (two-tailed permutation testing, $p < 0.05$). (**b**) Evolution of the topographical maps of the averaged mean power difference over a selected frequency range between rebuffering and playback signals of subjects #4. The numbers above the maps indicate the elapsed time from the onset of rebuffering in ms. (**c**) Individual time limits of eight subjects on the axis each indicated by a black arrow

of a stimulus [51], while high-level perceptual decision-making and classification activities occurred at 150 ms [49, 51]. It can be inferred that parietal beta-band oscillations are correlated with negative emotion. Therefore, the time limits we calculate represent thresholds after which subjects experience negative emotions. These time points do not align with previous findings on neural perception or cognition. The subjects in our experiment are not required to perform any tasks related to rebuffering, ensuring that their reactions inferred from the EEG signals reflect natural emotions evoked by rebuffering.

Summary We have discovered the reaction mechanism of the human brain to video rebuffering. After the appearance of a freeze frame, it takes approximately 75 ms for the brain to receive the visual signal from the eyes. Then, the processing of this signal begins. After 75 ms, the brain completes the classification of the stimulus and recognizes that this frame, which differs from normal video frames, is a freeze frame. At this moment, the classification result deviates from the viewer's expectations, but the high-level cognitive processing of the stimulus does not commence. From 190 ms to 850 ms after the stimulus, based on individual cognitive speeds, negative emotions are aroused, as reflected by parietal beta-band oscillations.

6.2.2.3 Calibration of the Effect of Video Rebuffering on QoE

The results of the aforementioned experiment guide the modeling of the effect of video rebuffering on QoE and the resource allocation of the DASH system. If the rebuffering length is shorter than the cognitive time limit, users do not experience negative emotions related to video rebuffering. Therefore, transient rebuffering events should be allowed, which can help avoid longer rebuffering caused by insufficient bandwidth allocation for other users.

The prerequisite for such a strategy is a quantitative measurement of the effect of video rebuffering length on user experience. Since the time limit of each user is unknown to the DASH client, an alternative approach involves calculating the probability that a given rebuffering length does not evoke negative emotions. Therefore, we first estimate the distribution of individual time limits. Let Φ represent a random variable denoting the time limit of a user in the proposed DASH system and assume that the distribution of time limits in this system follows the distribution observed in the aforementioned experiment. Next, we derive the parameters of five common probability functions (i.e., normal, gamma, Poisson, exponential, and Rayleigh) from the measured data using maximum-likelihood estimation (MLE). Then, each of these probability functions is tested by evaluating the p value of rejecting the hypothesis that the measured data follows the distribution. Fig. 6.15 presents the five hypothesized cumulative distribution functions (CDFs) along with the empirical CDF. Among the five hypotheses, the normal distribution provides the best fit to the data, leading to the assumption of $\Phi \mathcal{N}(\mu, \sigma^2)$, where $\mu = 524$ and $\sigma = 174$ are the MLE parameters.

According to the results regarding human cognitive time limits, the calibration of the effects of video rebuffering on user experience can be represented by the probability that the rebuffering length is shorter than the time limit of users:

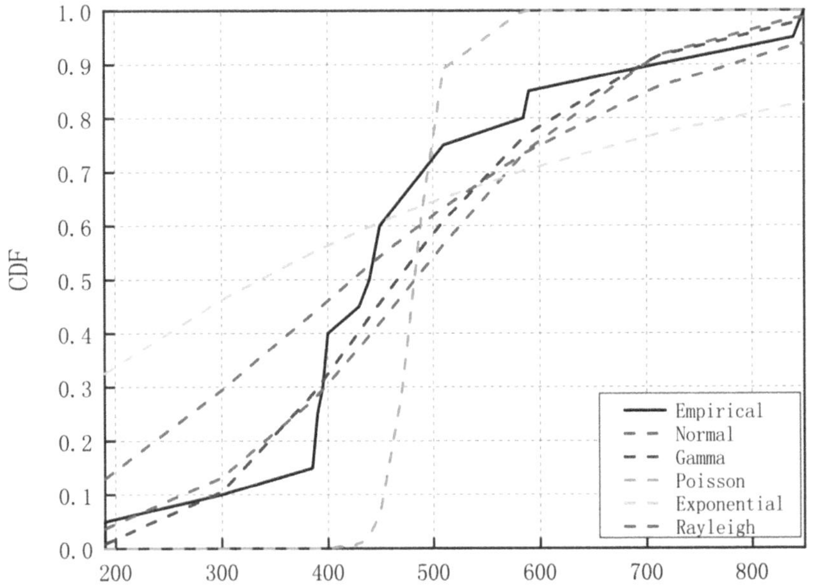

Fig. 6.15 CDF of the original data of individual time limits from eight subjects and the fitting curves of the CDFs of various distributions from the Kolmogorov-Smirnov test. The p values of rejecting the hypothesized distributions are 0.989, 0.967, 0.023, 0.115, and 0.560 for the normal, gamma, Poisson, exponential and Rayleigh distributions, respectively

$$PrRB_i \leq \Phi = 1 - Z(\frac{RB_i - \mu}{\sigma}), \tag{6.27}$$

where $Z(x) = \frac{1}{\sqrt{2\pi}} \int_{-\infty}^{x} e^{-t^2/2} dt$. The longer rebuffering lasts, the higher the likelihood that the user will experience negative emotions and the worse their experience will be.

6.2.3 Game Formulation

After deriving the analytic form of rebuffering length in the next slot and calibrating the effect of rebuffering length on user experience, we present the Stackelberg game formulation [22, 53] for the price-based bandwidth allocation scheme. The Stackelberg equilibrium of the proposed game is then investigated. In this section, we omit the time label for the sake of brevity, and the bandwidth allocation and BSI considered are for the next slot $\tau + 1$.

6.2.3.1 Game Formulation

We assume that the total bandwidth the DASH server can allocate to all users without incurring congestion is R_{total}, i.e.,

$$\sum_{u=1}^{N_U} R_u \leq R_{total}, \tag{6.28}$$

and the server charges each user a price p per unit of bandwidth. At the beginning of each frame, the server determines the optimal price and bandwidth allocation scheme based on the BSI received from all DASH clients. To ensure fairness and efficiency in the bandwidth allocation scheme, we set a unified price for all clients, and the bandwidth allocated to each client should be the optimum under the given price. This problem can be formulated as a Stackelberg game, which consists of a leader and several followers competing for resources.

Under the above game model, it can be easily observed that the server acts as the leader, aiming to maximize its revenue from selling bandwidth resources to the users. Mathematically, the revenue J_S can be expressed as:

$$J_S(p, \mathbf{R}) = \kappa \sum_{u=1}^{N_U} R_u, \tag{6.29}$$

where κ is the unit price of bandwidth and $\mathbf{R} = (R_1, R_2, \ldots, R_{N_U})$. The following price setting (PS) problem is solved to achieve the maximum revenue.

$$
\begin{aligned}
&(PS) \max_{\kappa} J_S(\kappa, R) \\
&s.t. \sum_{u=1}^{N_U} R_u \leq R_{total}.
\end{aligned}
\tag{6.30}
$$

Note that R_u actually depends on κ, because the amount of bandwidth that each user requests should maximize utility, which is:

$$
J_i(R_i, \kappa) = \lambda_i (PrRB_i \leq \Phi + \frac{B_i}{B_{max}}) - \kappa R_i \tag{6.31}
$$

where $\lambda_i > 0$ is a weight indicating the user's willingness to pay for services of high-quality services. Hence, the following request BR problem needs to be solved for each user i,

$$
\begin{aligned}
&(BR) \max_{\kappa} J_i(R_i, \kappa) \\
&s.t. 0 \leq R_i \leq R_{total}.
\end{aligned}
\tag{6.32}
$$

Combining problems PS and BR forms a Stackelberg game. The objective of this game is to find the Stackelberg equilibrium (SE) point, at which neither the leader (server) nor the followers (clients) have the incentive to deviate. The SE of the proposed game is investigated in the following subsection.

6.2.4 Stackelberg Equilibrium

For the proposed game, the SE is defined as follows: Let R_i be the solution to problem BR for user i given κ. Then, point R_i^* is a Nash equilibrium for user i if:

$$
R^*i = \arg\max_{R_i} J_i(R_i, \kappa). \tag{6.33}
$$

Next, for the server as the leader, the price should be optimized as follows:

$$
\kappa* = \arg\max_{\kappa} J_s(\kappa, R^*). \tag{6.34}
$$

The optimal price along with the optimal bandwidth request under this price is referred to as the SE.

Since the server collects the current BSI of all users at the start of each time slot, it can compute and predict the users' bandwidth requests at a given price. Therefore, problems PS and BR can be optimized by the server in a centralized manner, where PS represents the main problem and BR is the subproblem.

6.2.5 Algorithm to Solve the Stackelberg Equilibrium

The objective function of problem BR is nonconvex and involves a Z-function and a fractional function Therefore, its solution cannot be expressed analytically. Hence, we present an efficient searching algorithm followed by the algorithm for problem PS. Finally, we analyze the computational complexity of the proposed algorithms.

As discussed in Sect. 6.2.3, the objective function of problem BR represents the difference between two increasing functions. Therefore, we transform the problem into a canonical monotonic optimization (MO) form and solve it using a successive poly-block outer approximation algorithm [60]. Once again, we omit the script of the slot number τ.

We introduce an auxiliary variable

$$s = \kappa R_{total} - \kappa R_i \tag{6.35}$$

with a feasible interval

$$0 \leq s \leq \kappa R_{total}. \tag{6.36}$$

Using s, we equivalently transform problem BR into:

$$\begin{aligned} (BR - MO) \max_{R_i, s} J_i(R_i, s, \kappa) &= \lambda_i[1 - Z(\frac{RB_i - \mu}{\sigma}) + \frac{B_i}{B_{max}}] + s - \kappa R_{total} \\ s.t. (R_i, s) &\in \mathcal{X}, \end{aligned} \tag{6.37}$$

where set $\mathcal{X}$ is given by:

$$\mathcal{X} = \{(R_i, s) | (0, 0) \leq (R_i, s) \leq (R_{total}, \kappa R_{total} - \kappa R_i)\}. \tag{6.38}$$

Based on the monotonic optimization theory, $\mathcal{X}$ is a normal set. This problem can be formulated as monotonic optimization (MO) problem.

6.2.6 Experimental Results

In this section, we present the experimental results to validate the generalization ability of our EEG-based QoE model and the effectiveness of the proposed algorithm.

6.2.6.1 Validating the Results of the Distribution of User Time Limit via Parallel Experimentation

To validate that the results of our EEG experiments are not sensitive to the sampling of the subjects, we conduct experiments on an additional eight subjects, comprising four females and four males, following the same experimental design and data analysis methodology.

The frequency distribution of the EEG oscillations in the repeated experiment is similar to that of the previous experiment. Specifically, the frequency range of 15–24 Hz consistently exhibits a significant increase in signal during rebuffering events compared to playbacks (Fig. 6.16, one-tailed Wilcoxon signed-rank test, $p < 0.05$).The frequency distribution is also similar to the previous experiments, except that the number of subjects with a significant difference in the mean log power of the selected frequency range on P6 is 6 instead of 7. Therefore, we select ten channels that consistently show a significant difference in power during

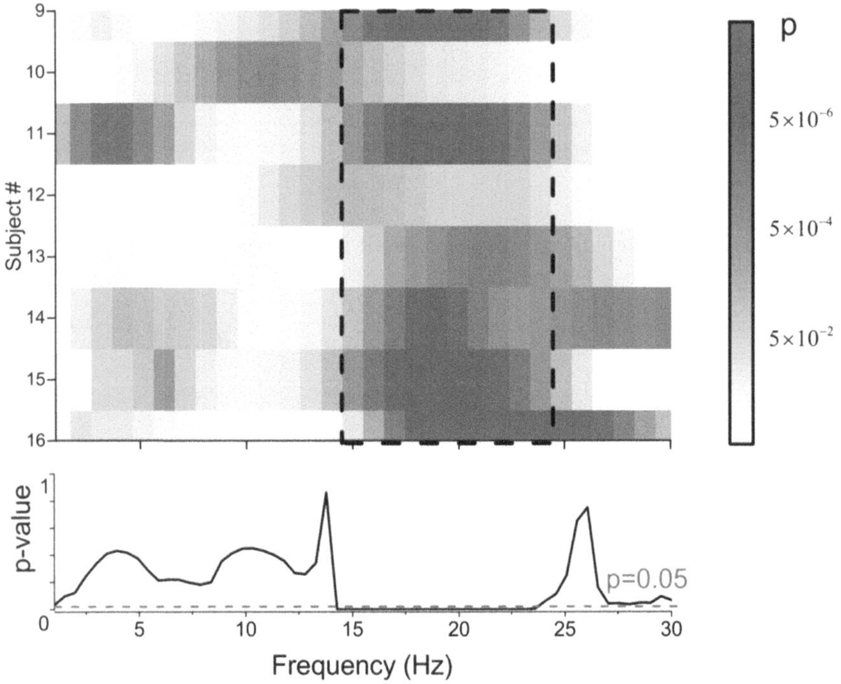

Fig. 6.16 EEG reactions to video rebuffering for the frequencies of another eight subjects. (Above) Significance levels of the difference (two-tailed t-test) between two conditions and (below) the statistical results of whether eight t-test p values at each frequency point are less than 0.05 (one-tailed Wilcoxon signedrank test).The dashed gray line in the bottom figure indicates the threshold $\kappa = 0.05$, whereas the dashed box in the top figure indicates the frequency band that chose according to the κ-values shown in the bottom curve

video rebuffering (Fig. 6.17). Moreover, the time limits of the onset of the captured components follow a normal distribution, with $\mu = 498$ ms and $\sigma = 131$ ms, which is similar to the previous experiment where $\mu = 524$ ms and $\sigma = 174$ ms. We also perform a permutation test on the two sets of time limits, and it turns out that there is no significant difference between their means (two-tailed permutation test, $\kappa = 0.9388$).

These brain reactions, characterized by similar frequency, channel range, and distribution of calculated time limits, validate the generalization ability of the EEG-based QoE model in relation to video rebuffering. The time required for a human to experience negative emotion due to video rebuffering follows a normal distribution. We further update the parameters in (18) by calculating the time limits of all 16 subjects, resulting in $\mu = 513$ ms and $\sigma = 164$.

6.2.6.2 Performance of the Proposed Bandwidth Allocation Algorithm

The simulation is conducted on an ns2 platform consisting of a DASH server, $N_U = 6$ DASH clients, bandwidth management, and a TCP/IP layer. The simulation time is set to 360 s per iteration, and all users are activated at the beginning of the simulation. Each client requests one sequence extracted from xiph.org, with a frame rate of 30 fps. The sequence is encoded into six profiles with rates φ ranging from 350 kbps to 3 Mbps, corresponding to common bit rate levels of 240 p, 360 p, 480 p, 720 p, and 1080 p. The chunk duration is set to $Tc = 2$ s and the period of bandwidth allocation is set to $Ts = 1$ s.

We adopt the rate-distortion-based algorithm (RDA) in all clients due to its low complexity. The buffer size is set to $Bmax = 30$ s. The available bandwidth at the

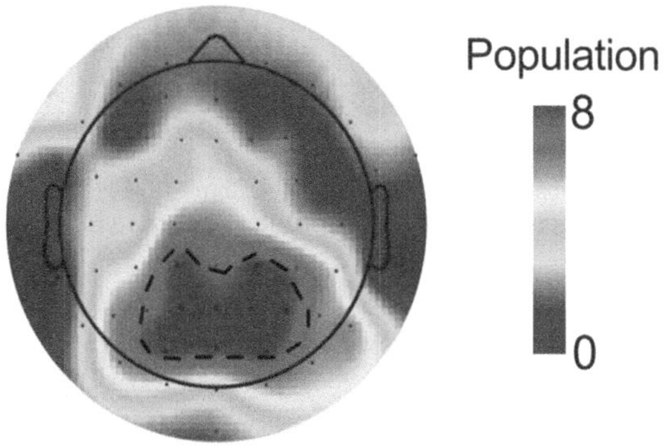

Fig. 6.17 EEG reactions to video rebuffering in channels of another eight subjects. topographical maps of the number of subjects who show a significant difference in mean log power in the frequency range from 15 to 24 Hz (two-tailed t-test, $\kappa < 0.0005$) at each channel. The dashed circle surrounds the channels with a population of $n7$. The selected channels are Pz, 1–5, 8; POz, 3–4, 7–8

server was set to $Rtotal = 5$ Mbps. The quality of the first chunk downloaded by each client is set to the lowest profile, and the playback of each client begins as soon as the first chunk is downloaded.

Performance in terms of QoE Next, we compare the distribution of the numbers and lengths of rebuffering events experienced by the users of the two algorithms over 100 iterations. As shown in Fig. 6.18, the RAJPBA algorithm provides users with significantly fewer rebuffering events compared to the PBBA and PQoE algorithms. In contrast, the PBBA and PQoE algorithms are more likely to cause long rebuffering events, resulting in a significant degradation in QoE. The distribution of the bit rates of chunks requested and displayed by the clients varies substantially between PBBA and RAJPBA, while the distribution varies slightly between RAJPBA and PQoE. In summary, the users have a lower probability of experiencing lengthy rebuffering when the RAJPBA algorithm is applied in the system, and the degradation in video bit rate is smaller compared to PQoE.

Pricing and QoE performance under varied available bandwidth We further increase the total available bandwidth from 5 Mbps to 20 Mbps to observe the difference in performances from the server side and the client side. Surprisingly, as shown in Fig. 6.19a, the optimal unit price of the bandwidth increase with the total available bandwidth, which is contrary to the rule of thumb that a seller will reduce the price when there is a more sufficient supply of goods. This phenomenon is due to the complicated relationship between the optimal bandwidth requirement $R{*}_i$ and κ. In different buffer states, $R{*}_i$ can increase, decrease, or remain unchanged

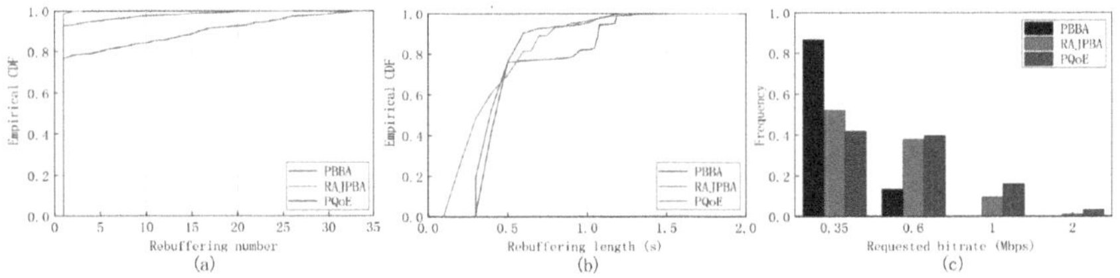

Fig. 6.18 Comparison of the performances of three algorithms: (**a**) CDFs of rebuffering numbers, (**b**) CDFs of rebuffering length, and (**c**) frequency distributions of requested bit rate

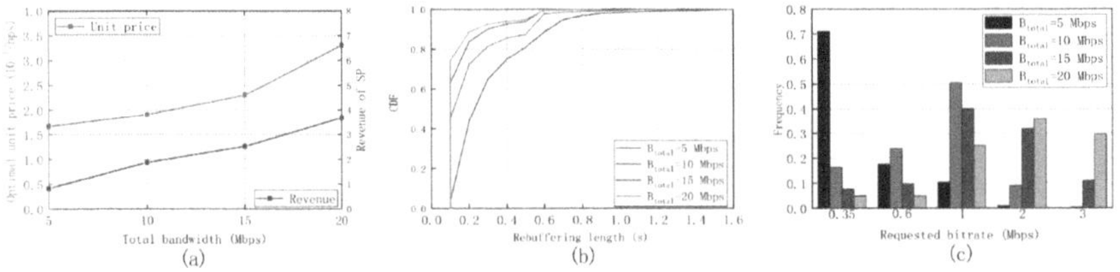

Fig. 6.19 System performance under varied total bandwidths: (**a**) optimal unit price and revenue of DASH server, (**b**) CDFs of rebuffering length, and (**c**) frequency distribution of requested bit rate

when p increases. As more available bandwidth enlarges the solution space of problem BR, the server needs to search for more prices. For this example, an increased p leads to both an increase and decrease in some $R*_i$, but the sum of the bandwidth requirement and the total revenue increases. In addition, it is observed from Fig. 6.19b and c that users' experience in terms of rebuffering length and video chunk bit rate improves with total available bandwidth.

References

1. Agustsson, E., Tschannen, M., Mentzer, F., Timofte, R., Gool, L.V.: Generative adversarial networks for extreme learned image compression. In: International Conference on Computer Vision (ICCV), pp. 221–231 (2019)
2. Akbari, M., Liang, J., Han, J.: Dsslic: Deep semantic segmentation-based layered image compression. In: 2019–2019 IEEE International Conference on Acoustics, Speech, and Signal Processing (ICASSP), pp. 2042–2046. IEEE, Piscataway (2019)
3. Androulidakis, A.G., Doyle, L.M., Yarrow, K., Litvak, V., Gilbertson, T.P., Brown, P.: Anticipatory changes in beta synchrony in the human corticospinal system and associated improvements in task performance. Eur. J. Neurosci. **25**(12), 3758–3765 (2007)
4. Arndt, S., Antons, J.N., Schleicher, R., Möller, S., Curio, G.: Using electroencephalography to measure perceived video quality. IEEE J. Sel. Topics Signal Process. **8**(3), 366–376 (2014)
5. Arndt, S., Brunnström, K., Cheng, E., Engelke, U., Möller, S., Antons, J.N.: Review on using physiology in quality of experience. In: Human Vision and Electronic Imaging 2016, HVEI 2016, 14 February 2016 through 18 February 2016, San Francisco; United States, pp. 197–205. The Society for Imaging Science and Technology (2016)
6. Ballé, J., Laparra, V., Simoncelli, E.: End-to-end optimized image compression. In: 5th International Conference on Learning Representations, ICLR 2017 (2017)
7. Bekisz, M., Wróbel, A.: 20 hz rhythm of activity in the visual system of perceiving cat. Acta Neurobiol. Exp. **53**, 175–175 (1993)
8. Bińkowski, M., Sutherland, D.J., Arbel, M., Gretton, A.: Demystifying MMD GANs (2018). arXiv preprint arXiv:1801.01401
9. Bjøntegaard, G.: Calculation of average PSNR differences between RD-curves. In: ITU-T VCEG-M33 (2001)
10. Blau, Y., Michaeli, T.: The perception-distortion tradeoff. In: IEEE Conference on Computer Vision and Pattern Recognition (CVPR), pp. 6228–6237 (2018)
11. BPG Image format (2018). https://bellard.org/bpg/
12. BT, R.I.R.: Methodology for the subjective assessment of the quality of television pictures. International Telecommunication Union (2002)
13. Cheng, Z., Sun, H., Takeuchi, M., Katto, J.: Learned image compression with discretized gaussian mixture likelihoods and attention modules. In: IEEE Conference on Computer Vision and Pattern Recognition (CVPR), pp. 7936–7945 (2020)
14. Cicalò, S., Changuel, N., Tralli, V., Sayadi, B., Faucheux, F., Kerboeuf, S.: Improving qoe and fairness in http adaptive streaming over lte network. IEEE Trans. Circuits Syst. Video Technol. **26**(12), 2284–2298 (2015)
15. Cisco: Cisco visual networking index: global mobile data traffic forecast update 2017–2022 (2017)
16. Cohen, M.X.: Analyzing neural time series data: theory and practice. MIT Press, Cambridge (2014)
17. Cordts, M., Omran, M., Ramos, S., Rehfeld, T., Enzweiler, M., Benenson, R., Franke, U., Roth, S., Schiele, B.: The cityscapes dataset for semantic urban scene understanding. In: Conference on Computer Vision and Pattern Recognition (CVPR), pp. 3213–3223 (2016)

18. Farsad, N., Rao, M., Goldsmith, A.: Deep learning for joint source-channel coding of text. In: 2018 EEE International Conference on Acoustics, Speech and Signal Processing (ICASSP), pp. 2326–2330 (2018). https://doi.org/10.1109/ICASSP.2018.8461983
19. Gola, M., Magnuski, M., Szumska, I., Wróbel, A.: Eeg beta band activity is related to attention and attentional deficits in the visual performance of elderly subjects. Int. J. Psychophysiol. **89**(3), 334–341 (2013)
20. Gong, R., Liu, X., Jiang, S., Li, T., Yan, J.: Differentiable soft quantization: bridging full-precision and low-bit neural networks. In: IEEE International Conference on Computer Vision (ICCV) (2019)
21. Goodfellow, I., Pouget-Abadie, J., Mirza, M., Xu, B., Warde-Farley, D., Ozair, S., Courville, A., Bengio, Y.: Generative adversarial nets. Adv. Neural Inform. Process. Syst. **27** (2014)
22. Han, Z., Niyato, D., Saad, W., Başar, T., Hjørungnes, A.: Game theory in wireless and communication networks: theory, models, and applications. Cambridge University Press, Cambridge (2012)
23. He, K., Gkioxari, G., Dollar, P., Girshick, R.: Mask R-CNN. In: Proceedings of the IEEE International Conference on Computer Vision (ICCV) (2017)
24. Heusel, M., Ramsauer, H., Unterthiner, T., Nessler, B., Hochreiter, S.: GANs trained by a two time-scale update rule converge to a local nash equilibrium. Neural Information Processing Systems (NIPS), pp. 6629–6640 (2017)
25. Hu, Y., Yang, W., Liu, J.: Coarse-to-fine hyper-prior modeling for learned image compression. In: AAAI (2020)
26. Installations, T., Line, L.: Subjective video quality assessment methods for multimedia applications. Networks **910**(37), 5 (1999)
27. Johnston, N., Vincent, D., Minnen, D., Covell, M., Singh, S., Chinen, T., Jin Hwang, S., Shor, J., Toderici, G.: Improved lossy image compression with priming and spatially adaptive bit rates for recurrent networks. In: IEEE Conference on Computer Vision and Pattern Recognition (CVPR), pp. 4385–4393 (2018)
28. Joseph, V., de Veciana, G.: Nova: Qoe-driven optimization of dash-based video delivery in networks. In: IEEE INFOCOM 2014—IEEE Conference on Computer Communications, pp. 82–90 (2014). https://doi.org/10.1109/INFOCOM.2014.6847927
29. Kroupi, E., Hanhart, P., Lee, J.S., Rerabek, M., Ebrahimi, T.: Eeg correlates during video quality perception. In: 2014 22nd European Signal Processing Conference (EUSIPCO), pp. 2135–2139. IEEE, Piscataway (2014)
30. Larsen, R.J., Marx, M.L.: Introduction to Mathematical Statistics and Its Applications: Pearson New International Edition. Pearson Higher Ed (2013)
31. Lee, J., Cho, S., Beack, S.K.: Context-adaptive entropy model for end-to-end optimized image compression. In: 7th International Conference on Learning Representations, ICLR (2019)
32. Li, X., Li, X., Luo, Y.J.: Differential influences of negative emotion on spatial and verbal working memory: Evidence from event-related potential and source current density analysis. Neuroreport **17**(14), 1555–1559 (2006)
33. Li, X., Shi, J., Chen, Z.: Task-driven semantic coding via reinforcement learning. IEEE Trans. Image Process. **30**, 6307–6320 (2021). https://doi.org/10.1109/TIP.2021.3091909
34. Lu, K., Li, R., Chen, X., Zhao, Z., Zhang, H.: Reinforcement learning-powered semantic communication via semantic similarity (2021). arXiv preprint arXiv:2108.12121
35. Lu, K., Li, R., Chen, X., Zhao, Z., Zhang, H.: Reinforcement learning-powered semantic communication via semantic similarity (2021). CoRR abs/2108.12121. https://arxiv.org/abs/2108.12121
36. Mentzer, F., Toderici, G., Tschannen, M., Agustsson, E.: High-fidelity generative image compression. Advances in Neural Information Processing Systems (2020)
37. Minnen, D., Balle, J., Toderici, G.: Joint autoregressive and hierarchical priors for learned image compression. In: Advances in Neural Information Processing Systems, vol. 31 (2018)
38. Mustafa, M., Guthe, S., Magnor, M.: Single-trial eeg classification of artifacts in videos. ACM Trans. Appl. Percep. **9**(3), 1–15 (2012)
39. Nitschke, W.H.J.B.: The puzzle of regional brain activity in and anxiety: the importance of subtypes and comorbidity. Cogn. Emotion **12**(3), 421–447 (1998)

40. Park, T., Liu, M.Y., Wang, T.C., Zhu, J.Y.: Semantic image synthesis with spatially-adaptive normalization. In: IEEE Conference on Computer Vision and Pattern Recognition (CVPR), pp. 2337–2346 (2019)
41. Qin, Z., Tao, X., Lu, J., Li, G.Y.: Semantic communications: Principles and challenges (2022). CoRR abs/2201.01389. https://arxiv.org/abs/2201.01389
42. Rabbani, M.: JPEG2000: image compression fundamentals, standards and practice. J. Electron. Imaging **11**(2), 286 (2002)
43. Ramamurthi, V., Oyman, O.: Video-qoe aware radio resource allocation for http adaptive streaming. In: 2014 IEEE International Conference on Communications (ICC), pp. 1076–1081. IEEE, Piscataway (2014)
44. Rippel, O., Bourdev, L.: Real-time adaptive image compression. In: Precup, D., Teh, Y.W. (eds.) 34th International Conference on Machine Learning (ICML), Proceedings of Machine Learning Research, vol. 70, pp. 2922–2930. PMLR (2017)
45. Scholler, S., Bosse, S., Treder, M.S., Blankertz, B., Curio, G., Muller, K.R., Wiegand, T.: Toward a direct measure of video quality perception using EEG. IEEE Trans, Image Process. **21**(5), 2619–2629 (2012)
46. Schutter, D.J., Putman, P., Hermans, E., van Honk, J.: Parietal electroencephalogram beta asymmetry and selective attention to angry facial expressions in healthy human subjects. Neurosci. Lett. **314**(1–2), 13–16 (2001)
47. Song, M., Choi, J., Han, B.: Variable-rate deep image compression through spatially-adaptive feature transform. In: International Conference on Computer Vision (2021)
48. Theis, L., Shi, W., Cunningham, A., Huszár, F.: Lossy image compression with compressive autoencoders (2017). arXiv preprint arXiv:1703.00395
49. Thorpe, S., Fize, D., Marlot, C.: Speed of processing in the human visual system. Nature **381**(6582), 520–522 (1996)
50. Toderici, G., Vincent, D., Johnston, N., Hwang, S.J., Minnen, D., Shor, J., Covell, M.: Full resolution image compression with recurrent neural networks. In: IEEE Conference on Computer Vision and Pattern Recognition (CVPR), vol. 2017, pp. 5435 – 5443 (2017)
51. VanRullen, R., Thorpe, S.J.: The time course of visual processing: from early perception to decision-making. J. Cogn. Neurosci. **13**(4), 454–461 (2001)
52. Wallace, G.K.: The JPEG still picture compression standard. IEEE Trans. Consum. Electron. **38**(1), xviii–xxxiv (1992)
53. Wang, Y., Wang, X., Wang, L.: Low-complexity stackelberg game approach for energy-efficient resource allocation in heterogeneous networks. IEEE Commun. Lett. **18**(11), 2011–2014 (2014)
54. Weng, Z., Qin, Z., Li, G.Y.: Semantic communications for speech signals. In: IEEE International Conference on Communications (ICC) (2020)
55. Xie, H., Qin, Z.: A lite distributed semantic communication system for internet of things. IEEE J. Sel. Areas Commun. **39**(1), 142–153 (2020)
56. Xie, H., Qin, Z., Li, G.Y., Juang, B.H.: Deep learning enabled semantic communication systems. IEEE Trans. Signal Process. **69**, 2663–2675 (2021)
57. Yin, X., Jindal, A., Sekar, V., Sinopoli, B.: A control-theoretic approach for dynamic adaptive video streaming over http. In: Proceedings of the 2015 ACM Conference on Special Interest Group on Data Communication, pp. 325–338 (2015)
58. Zahran, A.H., Quinlan, J.J., Ramakrishnan, K., Sreenan, C.J.: Sap: Stall-aware pacing for improved dash video experience in cellular networks. In: Proceedings of the 8th ACM on Multimedia Systems Conference, pp. 13–26 (2017)
59. Zhang, Y., Chen, Y., Bressler, S.L., Ding, M.: Response preparation and inhibition: the role of the cortical sensorimotor beta rhythm. Neuroscience **156**(1), 238–246 (2008)
60. Zhang, Y.J.A., Qian, L., Huang, J., et al.: Monotonic optimization in communication and networking systems. Found. Trends® Netw. **7**(1), 1–75 (2013)
61. Zhao, H., Shi, J., Qi, X., Wang, X., Jia, J.: Pyramid scene parsing network. In: Proceedings of the IEEE Conference on Computer Vision and Pattern Recognition (CVPR), pp. 2881–2890 (2017)

Chapter 7
Cloud-Edge-End Intelligent Coordination and Computing

7.1 Overview of Cloud Edge Collaboration

With the popularity of the Internet, cloud computing has been continuously developed and evolved for decades, playing its strong resource service capability in various applications and gradually becoming the mainstream computing paradigm. In order to solve the problem of cloud computing away from the data sources and user terminals, edge computing began to rise, and now, sinking the capability of cloud computing to the edge has become an important development trend. However, edge computing is often limited by the computing resources of the edge platform. In order to combine the advantages of cloud computing and edge computing together, cloud-edge collaborative technology came into being. With the development of technology, cloud-edge collaborative technology will inevitably play an increasingly important role in the future Internet industry. This chapter first starts with the introduction of cloud computing and edge computing, then analyzes and discusses the cloud-edge collaborative technology from all aspects, and finally ends with a summary and challenges in the research of cloud-edge collaboration technology.

7.1.1 Cloud Computing and Edge Computing

Cloud computing is a computing model which can conveniently provide the needed resources (e.g., computing resources, storage, information services, etc.) from a configurable resource-sharing pool through a network anytime and anywhere according to the user's needs, and these resources can be quickly supplied and released, reducing the workload of the user in managing resources and interactions with service providers to the least. Through cloud computing services, any individuals or groups can use their desired computing resources and dispel

X. Tao et al., *Wireless Multimedia Computational Communications*,
Wireless Networks, https://doi.org/10.1007/978-3-031-64155-8_7

the difficulty of building and maintaining servers and other infrastructure, thus concentrating on their own business. Cloud service providers can also manage and schedule computing resources uniformly, improving resource utilization efficiency and maximizing their own benefit.

The origin of cloud computing technology can be traced back to 1959 when Strachey raised virtualization technology in his paper [27], virtualizing computer hardware resources and solving the problem of interaction between hardware resources, setting up a significant technical foundation for cloud computing. In the late 1990s when the Internet was globally popular, more and more users were facing issues of complex tasks that were hard to handle by a single computer; thus, grid computing [12] is proposed to solve this problem, which fully utilizes idle computing resources of the heterogeneous computing devices in WAN, forming them into a cooperative computing cluster to solve large-scale computational problems. Grid computing technology has realized a computing service model based on networks, thus further promoting the development of the cloud computing model. In 2006, Google formally proposed the concept of cloud computing at its search engine conference, and subsequently, Amazon, Microsoft, and other firms carried out relevant research and launched their own cloud computing services. Nowadays, cloud computing technology has become mature and popular, and services of cloud computing such as Amazon AWS, Microsoft Azure, and Google Cloud are providing countless users with powerful computing resources, storage, etc.

The three most common service types of cloud computing currently are infrastructure as a service (IaaS), platform as a service (PaaS), and software as a service (SaaS) [14].

(1) IaaS: It provides customers with basic infrastructure such as servers, virtual machines, storage spaces, and network devices, such as Amazon AWS and Google Cloud Engine.
(2) PaaS: It offers a software development platform to customers, and they can develop, deploy, run, and maintain applications in its integrated environment, such as Google App Engine and Microsoft Azure.
(3) SaaS: It provides customers with application software functional service, and they can access the client interface through the network over various devices to use the service, such as Google Apps and Microsoft Office 365.

Cloud computing integrates software and hardware resources and offers services through networks, which is characterized by broad network access, rapid elastic scaling, self-service on demand, resource pool, and service metering. Apart from the variety of data-processing applications, cloud computing provides a superior platform for emerging technologies such as machine learning and big data so that researchers can concentrate on algorithms and data itself, instead of being constrained by software and hardware.

In addition to the advantages of cloud computing with strong resource-service capacity, drawbacks are gradually exposed due to new applications such as the Internet of Things (IoT), such as long end-to-end latency which cannot meet the requirements of timely sensitive applications, increasing the amount of data transfer which occupies a great amount of network bandwidth, as well as data security and privacy issues when transmitting all data to the cloud for processing.

To the issues above of cloud computing, research on constructing computation platforms at the edge of the network is widely popular, and edge computing technology has been rising rapidly in recent years. Edge computing is a new type of computing model that executes calculations at the edge of the network [25, 26]. It provides cloud services and an IT environment to application developers and service providers at the edge of the network; the goal is to provide computation, storage, and network for data input of users as closely as possible.

Corresponding to cloud computing service models, edge computing has mainly formed three service models: EC-IaaS, EC-PaaS, and EC-SaaS, which extend the service functions of the cloud side to the edge. Due to the geographical advantages of edge computing nodes, the end-to-end latency from user end to user end is reduced, and edge computing has distinct advantages over cloud computing in terms of application with high real-time requirements while reducing the bandwidth occupancy and improving data security and privacy.

On the other hand, as the name suggests, edge platforms are located at the edge of the network, and their wide distribution makes it difficult to have centralized management, bringing huge technical and financial difficulty. Also, there are huge differences in soft-hardware of each edge platform.

Nowadays, the era of the Internet of Everything has come. The number of user equipment connected to the Internet continues to grow, and the Internet of Things has become a new explosion point for the Internet. According to Transforma Insights' report, by 2022, there will be 13.4 billion Internet of Things devices, and it is estimated to increase to 30.5 billion by 2030 [17]. The traditional cloud computing architecture cannot meet this explosive mass data computing demand. It is an important development trend to sink the cloud computing capability to the edge and equipment side and to deliver, operate, and control uniformly through the center. In addition, the gradual maturity of 5G communication technology has greatly improved the network data transmission speed, which will also promote all kinds of applications to offload computing tasks to the cloud-edge collaboration platform for completion. Therefore, the current computing platform has been gradually transferred from the cloud stage to the cloud-edge collaboration stage.

Edge computing can be seen as an extension of cloud computing. The characteristics of the two are complementary: cloud computing can grasp the overall situation, process a large number of data and carry out in-depth analysis, and play an important role in scenarios such as non-real-time data processing; edge computing focuses on the local and better plays a role in the small-scale, real-time intelligent analysis. In conclusion, cloud-edge collaboration technology is committed to the complementary and coordinated development of cloud computing and edge computing, providing an important technical foundation for further development of the Internet industry as well as the intelligent and digital transformation of industries such as industrial production, transportation, and medical care.

7.1.2 Research Status of Cloud-Edge Collaboration

In the previous section, the concepts, technology development, and characteristics of cloud computing and edge computing were introduced. In this section, the collaboration technologies at different layers of cloud-edge will be analyzed and discussed. The primary issue of cloud-edge collaboration (cloud-edge-end collaboration) is how to collaborate between cloud, edge, and device, the concepts and characteristics of which are summarized briefly as follows:

- Cloud refers to the central nodes of traditional cloud computing, with the strongest computing and storage capabilities, generally located in data centers remote from geographical locations.
- Edge refers to the nodes built on the edge of the network, which is a medium between the cloud platform and the user equipment, with less computing resources than data centers, but geographically close to data sources and users.
- End refers to the user equipment, such as mobile phones, smart home appliances, various sensors, and cameras. User devices have high heterogeneity and limited computing capability, and mobile phones, etc. are greatly restricted by energy consumption.

Specifically, according to the whitepaper released by the Edge Computing Alliance, cloud-edge collaboration technologies mainly involve the collaboration between cloud and edge IaaS, PaaS, and SaaS [7], and therefore this section analyzes and discusses the research contents and status at the three levels respectively.

7.1.2.1 Cloud-Edge Collaboration at Infrastructure Level

Cloud-edge collaboration at infrastructure level mainly refers to the resource collaboration between cloud IaaS and EC-IaaS in terms of computing, network, and storage. One core issue of resource collaboration is resource scheduling. Compared with the cloud platform that has centralized vast resources, the computation, storage, and other resources at the edge nodes are relatively limited. When the application task has high resource requirements or the same node needs to serve multiple user requests at the same time, overflow problems may occur. Provisioning more capacities on the edge nodes would help reduce such problems, but seriously impairs the efficiency of cloud resource utilization during off-peak hours [2]. Thus, it is necessary to improve performance through efficient scheduling of cloud-edge platform resources. Li et al. [22] proposed the hierarchical distributed cloud service network model composed of Access Cloud, Distributed Micro Cloud, and Core Cloud established a resource description model based on the resource graph and hierarchical resource vector and proposed a particle swarm optimization–based multi-node system resource graph scheduling and allocation algorithm VCE-PSO, which solves the problem of resource shortage at the edge end by scheduling the resources in the cloud data center to the edge server and improve the utilization of cloud resources. If the flat set of the edge cloud server is used directly to provide service to the mobile users, due to the limited computing resources of the edge cloud server, the cloud access latency of the users will be very high at peak load. To solve this problem, Tong et al. [30] proposed to design the edge cloud server as a tree hierarchical structure, aggregate the peak load beyond the lower edge cloud server, and use the resources of the nodes at a higher layer to provide services, so that the edge can serve a larger peak load under the same computing resources and propose a workload placement algorithm to ensure the efficient use of computing resources of different layers in the edge cloud and minimize the average program execution delay in the edge cloud.

For applications that are sensitive to delay, such as intelligent transportation and driverless vehicles, it is necessary to reduce the server response time with scheduling. Tan et al. [28] proposed the OnDisc online job distribution and scheduling algorithm based on edge cloud according to the delay sensitivity of each task and assumed that tasks are published in any order and at any time by mobile devices, and the processing time of each job on different servers is not related, which is more in line with the real scene. It minimizes the weighted response time of all jobs.

Another important problem of resource cooperation is the cost problem. Especially in the process of commercializing cloud-edge collaboration technology, the cost problem is one that service providers need to consider carefully. Li et al. [20] proposed a resource management strategy to satisfy the edge cloud workload and minimize the cost of renting nodes by considering the idle renting nodes in the resource management strategy and proposed a dynamic replica allocation strategy and replica consistency preservation strategy. At the same time, it meets user experience and ensures data consistency, while reducing storage overhead.

7.1.2.2 Cloud-Edge Collaboration at Platform Level

Cloud-edge collaboration at the platform level can be further divided into data collaboration, intelligence collaboration, application management collaboration, etc.

1. **Data Collaboration**: Data collaboration involves operations such as data collection, processing, and storage on the cloud-edge platform. Usually, massive and heterogeneous data from terminal devices, sensors, and cameras are collected and transmitted to edge ends for preprocessing and then preprocessed relatively accurate, and complete data are transmitted to the cloud for further analysis and storage. Data preprocessing includes data cleaning, data integration, data reduction, etc. Data cleaning includes methods such as signal denoising and the elimination of outliers. Hodge, et al. [16] provided an overview of the out-liar technique and compared the motives, pros, and cons of various technologies. Outlier detection technology can be divided into three types: determining the outliers without prior data knowledge, similar to unsupervised clustering;

modeling normal and abnormal data, similar to supervised classification; and only modeling normal data, like semi-supervised recognition. It is concluded that there is no single universal outlier detection method, and the appropriate method should be selected according to the specific scenario and requirements. Carrizales et al. [4] implemented an efficient data preparation and retrieval approach for cloud storage, which mainly builds an index by hashing the data to identify and delete repeated data to reduce the cost of repeated data uploading and improve the efficiency of data preparation. Data aggregation refers to the dimensional reduction of datasets to avoid the curse of dimensionality. At present, principal component analysis (PCA) [9], sparse principal component analysis (SPCA) [37], factor analysis [19], etc. have been widely used for data dimension reduction. At present, deep learning is a popular data processing method on the cloud. Facebook, Apple, and Google use deep learning algorithms to analyze the large amounts of data they collect. Just as Apple's SIRI collects data from customers and analyzes user needs through deep learning, Google uses deep learning algorithms to realize Google translation, image and video search, and speech recognition applications. Jan et al. [18] compared various deep learning technologies used in big data analysis and categorized the big data tools used for specific scenarios. For large-scale full data, batch processing is usually conducted on the cloud computing platform using big data and deep learning technologies, while many scenarios require efficient processing of data streams and the latency brought by unloading to the cloud is often unacceptable. Therefore, usually edge platform is used for fast-stream data processing. Kuiper implemented an edge-based real-time streaming processing frame; furthermore, some scenarios require the batch streaming fusion processing of stored full-scale data and real-time generated stream data. Tyler Akidau et al. from Google proposed the Dataflow Model, which aims to provide a stream batch integrated data processing system [1], and for the calculation environment of cloud-edge collaboration and the characteristics of the edge end environment, Xinwei FU, et al. proposed the Edgewise streaming computing engine, which introduced an efficient operation scheduler to reduce the cost of operating system scheduling and reduce latency [13].

2. **Intelligent Collaboration**: Deep neural networks have been widely applied in high-level tasks such as computer vision, natural language processing, and optimization control. Nevertheless, due to the limited resources of computation storage at the edge, how to optimize the model and reduce the model size to deploy the deep neural network at the edge is a key issue. Deep neural network (DNN) partition and DNN compression are two common methods to achieve cloud-edge intelligent collaboration. Surat Teerapittayanon et al. [29] proposed a novel DNN architecture: BranchyNet, which is able to leverage the insight that many test samples can be correctly classified early and therefore does not need the later network layers, thus accelerating inference through the early exit of branches. Li et al. [21] used BranchyNet to train DNN models to achieve early exit and proposed a joint optimization algorithm to find the splitting points and exit points. Zhang et al. [35] formulated DNN partitioning as the min-cut

problem in the directed acyclic graph and proposed a two-stage solution for this problem. However, the model splitting requires frequent high-dimensional intermediate feature transmission between edge and cloud, thus resulting in increased latency. DNN compression is a mainstream model inference optimization technique. Han et al. [15] conducted pruning for unnecessary connections of the model and compressed the deep neural network with trained quantization and Huffman coding. Yang et al. [33] proposed an end-to-end DNN training framework, which transforms the DNN training formula into an optimization problem, and solved it by approximation algorithms. Zhang et al. [36] used an enhanced learning-based DNN compression method to generate lightweight models that can be deployed to the edge from heavyweight models, deployed the lightweight model to the edge and the heavyweight model in the cloud, and applied a supervised learning-based offloading strategy to determine when the samples are to be processed at the edge or in the cloud, compared to the cloud-only strategy; both inference latency and overall accuracy are improved.

3. **Application Management Collaboration**: The hardware on the edge platform holds high heterogeneousness, while Cloud-Native software architectures such as microservices, containers, and virtualization technologies can reduce application deployment and operation issues caused by hardware infrastructure differentiation [6]. Nowadays, the development of sophisticated applications using microservice technologies (e.g., Kubernetes, Apache Mesos, etc.) has become a trend among increasingly influential IT companies or application providers. The advantage of breaking down applications into several logically related but functionally independent microservices could bring high flexibility and reusability. With the help of microservice architecture [11], more complex applications will be easily developed using services in some specific orders. And with the help of container-based technologies, these microservice-based applications can be deployed on edge servers [8] easily. Xiong et al. [32] proposed an edge-computing architecture KubeEdge based on Kubernetes platform and demonstrated that Kubernetes can be used to manage remote edge nodes using the same application programming interface (API) and to deploy and manage applications on the edge. Deng et al. [8] focused on the scenarios of deploying microservice-based applications on edge service supply systems and modeled the microservice instance of servers as queuing nodes, proposing an approach to finding the optimal deployment plan while satisfying the application average response time requirements at a lower cost. As the computational resources at the edge node are limited when the computation resource at the edge is redundant, more tasks can be deployed and executed locally; however, when the edge resources are not sufficient, some tasks are migrated to the cloud to compensate for the performance losses, which is the application on-demand deployment. Zhang et al. [34] formally model the on-demand deployment problem and propose a graph-based algorithm to search for the total latency-minimizing deployment strategy in polynomial time and propose an event-driven deployment algorithm to reduce the online searching cost.

7.1.2.3 Cloud-Edge Collaboration at Software Level

In terms of software-level cloud-edge collaborative services, edge nodes implement some EC-SaaS services according to the cloud-side policies, realizing customer-oriented on-demand SaaS services through EC-SaaS-SaaS cooperation, while the cloud-side mainly provides the service distribution policies between cloud-side and edge nodes, as well as the SaaS service capability undertaken by the cloud-side.

Most existing micro-service deployment and scheduling studies only deal with static networks and fixed service requests and neglect the dynamic changes concerning the user mobility–related resource availability and service requests [3, 10] in the mobile edge computing environment. The user mobility would lead to frequent switching of the nearby edge clouds, thus increasing the service delay when the user leaves its service edge cloud. Some researchers use service migration to solve this problem, which involves whether, how, and where to migrate services when the user changes its location [23, 24]. However, these researches only focused on the migration of one individual service, while not considering the other deployments to serve the mobile user, and cannot fully utilize the dynamic network resources. To address this issue, Wang et al. [31] formulated the micro-service coordination problem without any prior knowledge as a Markov decision process, proposed a reinforcement learning (RL)-based online micro-service coordination algorithm to realize optimal decision-making, and developed an edge cloud-to-edge cloud micro-service coordination scheme with low costs to reduce the overall service delay. Chen et al. [5] found that the service deployment in the prior edge cloud hybrid environment used simple, generic heuristic methods and thus proposed a micro-service-based deployment problem (MSDP) based on the heterogeneity and dynamic characteristics in the edge cloud hybrid environment and designed a multiple buffer deep deterministic policy gradient algorithm (MB-DDPG) to solve MSDP, which is a reinforcement learning–based algorithm using multiple buffers in a memory pool to better learn the dynamic characteristics and ultimately form deployment policies with higher-level objectives, such as minimizing the average waiting time of all the requests in the IoT devices.

7.1.3 Conclusions and Research Challenges

As mentioned above, cloud computing distributes and provides shared hardware and software resources to users through the Internet so that computing resources can be purchased and used on demand, just like tap water and electricity. Edge computing is a distributed architecture that reduces the latency of responding to user requests by offloading tasks to the network edge, which is closer to the data sources and user terminals.

Although cloud computing can provide a large number of computing resources, there are problems such as high latency, limited network bandwidth, and privacy. Edge computing is beneficial to solve the above problems, but it is facing the problem of resources being seriously limited. Therefore, cloud-edge collaboration technology is needed to combine cloud computing and edge computing closely. Edge computing is for real-time, small-scale task processing and realizes efficient intelligent local service, while cloud computing is for long-term, large-scale task processing and performs global communications, scheduling, and control on edge nodes, overall business arrangement, and application life cycle management, thus forming a hierarchical cloud-edge collaborative platform. In order to meet the requirements of delay, bandwidth, privacy, and energy consumption of the data processing platform in complex application scenarios such as industrial Internet, intelligent transportation, and smart city, we need to give full play to the advantages of various platforms in the cloud-edge collaboration system and realize the collaboration between different platforms.

On the other hand, cloud-edge collaboration still faces many problems and challenges:

- Heterogeneity of software and hardware. Whether it is edge platforms or user devices, there is high heterogeneity in software and hardware, and their computing and communication resources are severely imbalanced. Moreover, the scale of the devices is very large, which poses a huge challenge to how to manage the resources of the cloud edge, and needs further research to realize the compatibility and coordination between heterogeneous devices.
- Task offloading problem. Cloud-edge collaboration is for various application scenarios, and various levels of platforms in the cloud-edge collaboration system have different characteristics of computing capabilities, geographical locations, and data security. How to realize the advantages of cloud-edge collaboration by offloading the tasks to the corresponding platforms in the most suitable granularity and the optimal strategy is a key problem of cloud-edge collaboration technology.
- Data security and privacy protection. Privacy protection technology has gone through decades of development and has successively put forward technologies such as homomorphic encryption, differential privacy, and federated learning. On the one hand, these technologies have their own limitations and technical bottlenecks. On the other hand, the era of big data has come and new application scenarios have put forward higher challenges to privacy protection technology. At present, the research on the data security of cloud-edge collaboration is still in its infancy [6].

7.2 End-Side Active Sensing and Computing

Under the booming tide of artificial intelligence, it becomes more and more important to develop end-side cloud collaborative intelligence. End-side intelligence can be applied to differentiated scenarios so that more critical information is retained. Its combination with cloud intelligence and collaborative computing is the best solution to implement the calculation with scene information. A typical example of computing on the side is the active vision camera. Different from passive video recording cameras, active visual cameras can make active perception and calculation adaptively according to the scene and process key information locally. Its framework involves the redesign of sensors and image signal processor (ISP) in the camera. This section takes the active vision camera frame designed by Liu Changsong's team at Tsinghua University as an example, to introduce the architecture and methods of end-side active sensing and computing.

7.2.1 Significance of Loading the Calculation on the End Side

7.2.1.1 Prosperity and Challenge of Cloud Intelligence

Cloud intelligence is booming, and computing has indeed changed the world, from DeepBlue, a computer that competed for chess with men, to DeepMind, a Go machine that successfully played against humans. The exponential growth of computing and storage capabilities described by Moore's Law, the massive data brought by the explosive growth of the Internet and the Internet of Things, and the rapid development of intelligent algorithms have greatly promoted the development of artificial intelligence from theory to implementation. We are becoming more and more familiar with the services that place algorithms in the cloud, such as face swiping and convenient payment, and cloud intelligence is getting smarter as it grows.

But cloud intelligence also faces many challenges. In the tide of open-source algorithms, it is difficult to achieve differentiated user experience and technological innovation. And its inefficient input-output ratio continues to be criticized. In addition, the space for cloud intelligence is limited. Many practical requirements cannot be addressed, such as image quality, network bandwidth, real-time performance, and reliability.

7.2.1.2 Growing Importance of Terminal Intelligence

Under the background of many demands and technological development, terminal intelligence arises at a historic moment and enjoys more and more popularity. It has high real-time and privacy protection abilities and saves resources.

Most end-side intelligence now simply puts the computation on the front end. However, it can accomplish the more important ability of intelligent environment sensing. This will greatly alleviate the bottleneck problem in the recognition system, that is, image quality.

7.2.2 *Active Visual Camera System*

7.2.2.1 Framework of Active Visual Camera System

A video transmission system is a monitoring system for shooting, recording, and transmitting remote images and video sequences. The common video transmission systems aim to see and record the remote image, which has no high requirement for real-time intelligent computing. Figure 7.1 shows the processing of common video transmission systems, which includes sensors, ISPs, video compression, and transmission, communication protocols, and output. However, real-time intelligent sensing and computing require high technology in scenarios where timely warning, real-time interaction, and real-time control are required.

The combination of artificial intelligence technology and traditional video surveillance system is the current research trend. Traditional image recognition technology relies on the support of large hardware such as servers and adopts the call mode of cloud intelligence, that is, the video system is transferred to the cloud for complex computation like image processing. In practical application, cloud intelligent technology relies on cloud calls to realize online image recognition. Cloud intelligence also has some drawbacks, such as difficulty in differentiation, low input-output ratio, inability to solve many practical problems, and limited space for development.

Terminal intelligence is becoming more and more important today. Different from cloud intelligence, it moves smart technology forward to capture video objects and sense environments smartly at the same time. However, the existing end-side solutions are still simple and rough and do not integrate intelligent technology with sensing, image processing, compression, and other steps. Intelligent environment sensing is more important, where image quality is the bottleneck of the recognition system. This project intends to propose an intelligent camera technology scheme to achieve very low-delay video target capture and transmission. As shown in Fig. 7.2, the intelligent algorithm is combined with a front-end sensor, ISP, and image compression technology for active image perception.

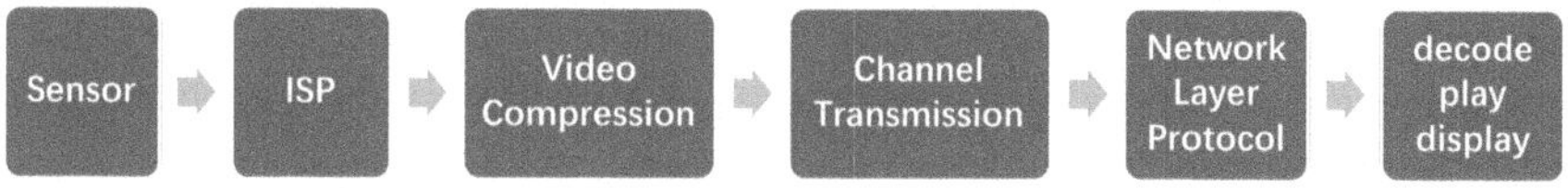

Fig. 7.1 Process diagram of common video transmission systems. System composition: sensor, ISP (input, output, frame buffer, and calculation), compression coding, channel transmission (network characteristics and distance), network layer protocol (packet loss retransmission), decoding, play buffer, and display

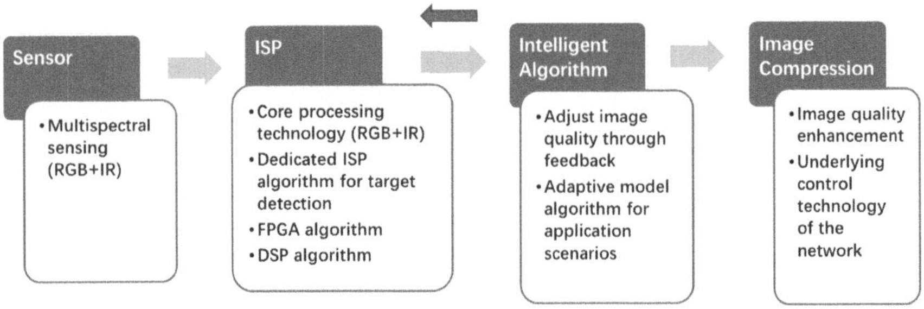

Fig. 7.2 Active image intelligent sensing system from Liu Changsong's research group at Tsinghua University. Smart algorithms are combined with technologies such as sensors, ISP, and image compression

7.2.2.2 Intelligent Sensing: A Solution for RGB-IR Sensors

IR channel data are of great significance for target detection, recognition, in vivo detection, fog penetration, and complex lighting scene processing. In order to synthesize the advantages of visible and near-infrared light, the general technical solution to the illumination problem takes dual light sources and the integration of visible and infrared light sources. There are mainly two schemes for integration: a dual-camera scheme, a combination of a visible camera and an infrared camera, which can obtain two kinds of pictures under visible and infrared light at the same time; a single-camera solution, a camera equipped with both visible and infrared light sensor. The latter requires an additional infrared (IR) cut to switch between visible and infrared light to obtain the corresponding illumination images. Under the existing technology, the application environment of the dual-camera scheme is complicated and inconvenient, requiring two cameras to work simultaneously and designing image processing algorithms for different cameras respectively. The single-camera switching scheme cannot get visible and infrared images at the same time, and the switching time may cause motion loss in imaging, resulting in inaccurate images. Figure 7.3 shows common compositions of two single cameras. The bottom one is a typical pattern with an IR cut; the top one is more advanced, a version of IR-cut free.

The active vision camera in this section uses a single camera scheme and a special RGB-IR multispectral sensing technology to overcome the above imaging defects. Figure 7.4 shows the general principles of the RGB-IR sensing system. After the camera receives light, two sets of light images are obtained through a specific RGB-IR color filter array (CFA) and processing block. This novel scheme is free from IR-cut and uses an IR component instead, which reduces cost and eliminates switching noise. Its imaging capability in white light is close to that of an ordinary camera, and it can achieve better night vision at night.

But the RGB-IR single-camera system is likely to distort the color of overlit pixels. The team from Tsinghua University perfectly solved this problem with a CPU-based processing scheme, FPGA hardware implementation, and unique DSP

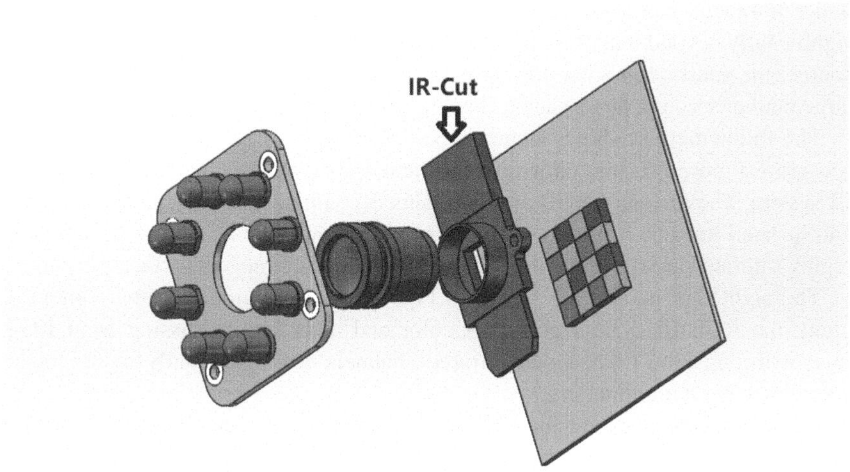

Fig. 7.3 Common camera compositions in the single-camera scheme. Classic camera (the bottom one) is equipped with IR-cut, and the advanced camera (the top one) is free of it

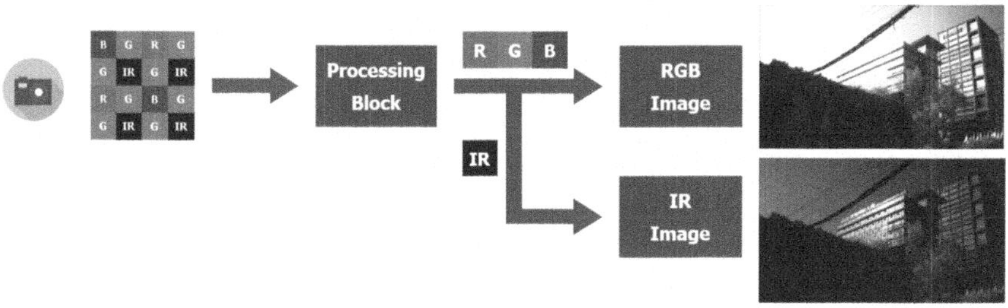

Fig. 7.4 General principles of RGB-IR sensing systems. Special color filter arrays are used. Visible and infrared images can be obtained in the final imaging results

software calculations. They designed advanced algorithms for the IR component: during the day, infrared channels are auxiliary, improving the dynamic range under bright light; at night, infrared is used for night vision replenishment to enhance brightness and maintain color. The four-channel multispectral imaging technology with a special CFA makes the collected information more abundant.

7.2.2.3 Advantages and Cases of Spectral Extension

There are many advantages to the unique RGB-IR sensor system. The collected IR channel data are useful for target detection, recognition, in vivo detection, fog penetration, and processing of complex lighting scenes. It has high image quality under luminous conditions and is free from IR-cut, so its application scene is significantly beyond that of traditional cameras. Without an additional infrared light source, the sensor can solve the problem of multispectrum acquisition and has a large market space in fields such as health, agriculture, and education.

The following part shows some cases of the system's performance in different scenarios. Figure 7.5 shows a wide dynamic range of images obtained by the RGB-IR system. The fusion of RGB and IR channels combines the advantages of visible and infrared images. It preserves the color of the image while adding detail to the highly saturated areas, increasing the brightness sensing range by a factor of three.

The night shot images are shown in Fig. 7.6. In an illuminated environment at night, the RGB-IR system preserves color and adds details missing from RGB images through the IR channel. infrared channels can significantly reduce noise when there is no illumination

Figure 7.7 suggests that the sensor system has great potential for application in driving scenarios. The introduction of infrared information makes the status

Fig. 7.5 The RGB-IR system increases the dynamic range of the images with the infrared channel. (**a**, **d**) RGB images. (**b**, **e**) IR images. (**c**, **f**) Images by RGB-IR system

Fig. 7.6 The RGB-IR system performs well at night in both illuminated and unilluminated conditions. (**a**, **d**) RGB images. (**b**, **e**) IR images. (**c**, **f**) Images by RGB-IR system. (**a**–**c**) Images at night with light. (**d**–**f**) Images at night without light

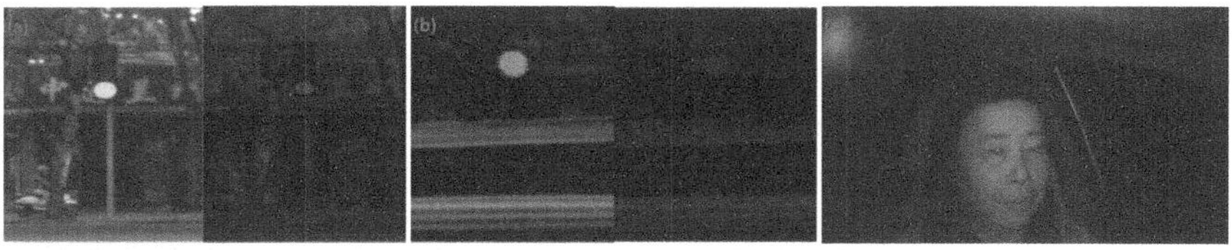

Fig. 7.7 Application case of RGB-IR system in driving scenario. (**a**, **b**) RGB and IR images of traffic lights. (**c**) Condition monitoring of the driver at night

Fig. 7.8 Comparison of fog penetration effect between ordinary color camera (**a**) and RGB-IR algorithm (**b**)

recognition of traffic lights more accurate. The system can also monitor the driver's body condition and maintain high-quality imaging at night. It can play a role in automatic or assisted driving with a wide dynamic range and excellent fog penetration and scene recognition ability.

Figures 7.8 shows that the RGB-IR sensing system has outstanding fog penetration capability. On foggy days, Tsinghua's intelligent algorithm produces images with enhanced contrast and increased detail compared to that of an ordinary color camera.

In Fig. 7.9, the RGB-IR sensing system is used for biometric recognition, in the case of hands. Depending on the depth of red, green, blue, and near-infrared penetration through the skin, the system can get four images of the palm with different biological features. Near-infrared (NIR) has the deepest depth and can be

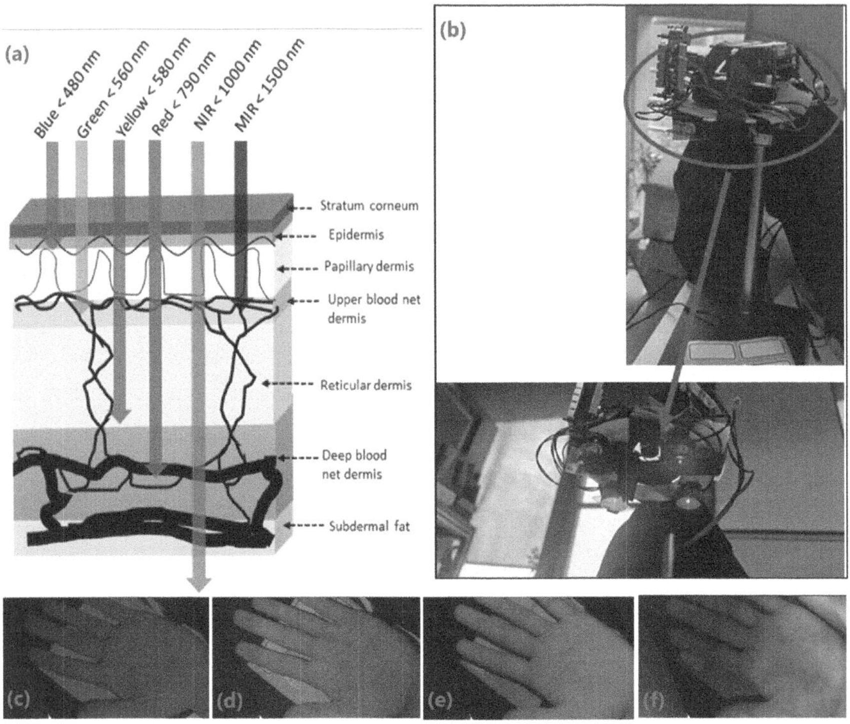

Fig. 7.9 The RGB-IR sensing system is used to identify the biometric features of hands. (**a**) Diagram of the penetration depth of different bands of light in the skin. (**b**) RGB-IR system that can receive RGB and IR channels at the same time. (**c**–**f**) Images of hand under blue, green, red, and near-infrared light

used for venous imaging and recognition. The system can also be applied to in vivo detection, with the rich physiological information of the four band images.

Authenticity authentication is another application scenario. For example, in Fig. 7.10, combining the imaging characteristics of banknotes under visible and infrared light will help to determine the authenticity of banknotes.

7.2.2.4 The ISP Technology Inside the Camera

The traditional RGB-IR method introduces IR-cut to physically filter RGB and IR channels. However, this scheme cannot obtain the fusion signals of RGB and IR at the same time, which is equivalent to switching two sensors controlled by a switch. Different from only RGB channels, when an IR channel is introduced, special processing schemes need to be redesigned for joint channels.

Figure 7.11 shows an example of a processing scheme. Additional steps are introduced to generate IR images and adjust RGB images, compared to the

Fig. 7.10 Banknote images in IR (**a**) and RGB (**b**)

processing flow when only RGB channels are available. The combination and complementarity of RGB and IR images will greatly improve the efficiency of face capture.

7.2.2.5 End-Side Computing: More Efficient Identification Method

Video image computing includes the underlying image processing of original signals from the sensor, face location, facial key point tracking, expression recog-

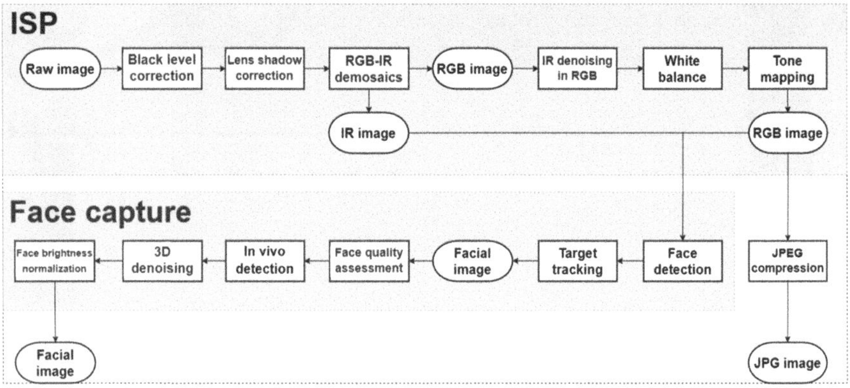

Fig. 7.11 Processing framework for RGB-IR joint channel

nition, and physiological signal extraction. All of these require a huge amount of computation and extremely high precision.

In this design scheme, these calculations will be left to the camera side, so as to ensure the minimum signal distortion, low delay, cost reduction and reduce the computing burden of the host, as opposed to the traditional way where images are compressed and then transferred to the host for processing.

Each pixel requires a large amount of the same processing by the ISP, with vector computation predominating. Convolutional neural network (CNN) is used to realize face location, tracking, and expression recognition, which involve a lot of convolutional operations. Physiological signal extraction also needs a lot of general computing support. To process these calculations simultaneously in real time, even a powerful computer mainframe requires CPU and GPU acceleration with strong calculating capacity.

There are two possible ways to implement these calculations on the camera side, through dedicated hardware (ASIC or FPGA) or general-purpose processors (CPU). The former has a high cost and a long development cycle, while the latter has limited computing power and is difficult to achieve optimal calculating efficiency. Heterogeneous computing is one of the effective solutions to the problem of computing power and efficiency of general-purpose processors.

Heterogeneous computing refers to the collaboration between CPU and GPU or CPU and other devices with different architectures like FPGA, DSP, and dedicated accelerators. Common programs are implemented on the CPU, but when large amounts of data need to be calculated, the CPU can be overloaded. We designed and configured other processing units capable of parallel processing: a neural network accelerator (NPU) for efficient processing of neural net-related operations and a parallel vector computing unit for ISP-related calculations. These processing units are linked together by high-speed buses, and each efficiently computes and processes the data in which they excel.

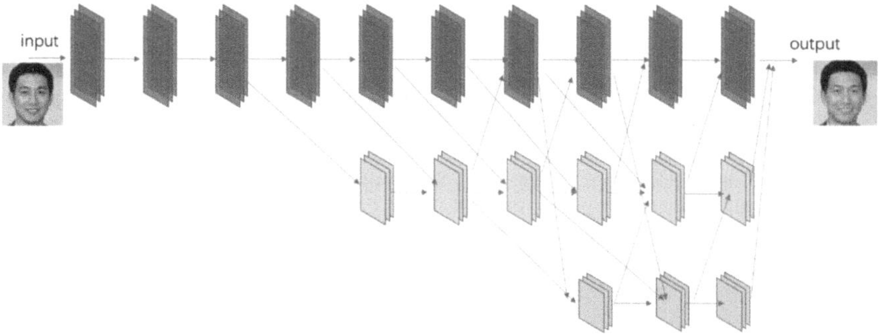

Fig. 7.12 Schematic diagram of a high-resolution subnetwork algorithm for facial keypoint detection

The difficulties of heterogeneous computing include the design and selection of hardware, the design of core algorithm based on special NPU, and the design and optimization of image processing algorithm based on parallel processing units. Taking the facial keypoint detection algorithm as an example, this system improved the detection method based on high-resolution Net (HRNet). The target of facial key point detection was to detect the position of N_s points in an input image I (size $W_{\text{image}} \times H_{\text{image}} \times 3$), which was realized by an improved high-resolution subnetwork algorithm, as shown in Fig. 7.12. By incrementally increasing the subnets from high resolution to low resolution, more stages were formed and multi-resolution subnets were connected in parallel. Then, through multiple multi-scale fusions, each representation from high resolution to low resolution received information repeatedly from other parallel representations; thus high-resolution representations with rich information were obtained. The problem of key point location detection was transformed into a task that estimates K_h heatmaps (each size $W'_{\text{image}} \times H'_{\text{image}}$), where the k_h-th heatmap represents the probability of the k_h-th key appearing at each location. Finally, 88 key points were located.

References

1. Akidau, T., Bradshaw, R., Chambers, C., Chernyak, S., Fernández-Moctezuma, R.J., Lax, R., McVeety, S., Mills, D., Perry, F., Schmidt, E., Whittle, S.: The dataflow model: a practical approach to balancing correctness, latency, and cost in massive-scale, unbounded, out-of-order data processing. Proc. VLDB Endowment **8**, 1792–1803 (2015)
2. Armbrust, M., Fox, A., Griffith, R., Joseph, A.D., Katz, R., Konwinski, A., Lee, G., Patterson, D., Rabkin, A., Stoica, I., et al.: A view of cloud computing. Commun. ACM **53**(4), 50–58 (2010)
3. Balalaie, A., Heydarnoori, A., Jamshidi, P.: Microservices architecture enables devops: migration to a cloud-native architecture. IEEE Softw. **33**(3), 42–52 (2016)
4. Carrizales, D., Sánchez-Gallegos, D.D., Reyes, H., González-Compeán, J., Morales-Sandoval, M., Carretero, J., Galaviz-Mosqueda, A.: A data preparation approach for cloud storage based

on containerized parallel patterns. In: Internet and Distributed Computing Systems: 12th International Conference, IDCS 2019, Naples, Italy, October 10–12, 2019, Proceedings 12, pp. 478–490. Springer, Berlin (2019)
5. Chen, L., Xu, Y., Lu, Z., Wu, J., Gai, K., Hung, P.C., Qiu, M.: IoT microservice deployment in edge-cloud hybrid environment using reinforcement learning. IEEE Internet Things J. **8**(16), 12610–12622 (2020)
6. Chen, Y.p., Liu, B., Lin, W.w., Cheng, H.W.: Survey of cloud-edge collaboration. Comput. Sci. **48**(3), 10 (2021)
7. computing Consortium, E., of Industrial Internet, A.: White paper on edge computing and cloud computing collaboration (2023). http://www.ecconsortium.org/Uploads/file/20190221/1550718911180625.pdf
8. Deng, S., Xiang, Z., Taheri, J., Khoshkholghi, M.A., Yin, J., Zomaya, A.Y., Dustdar, S.: Optimal application deployment in resource constrained distributed edges. IEEE Trans. Mobile Comput. **20**(5), 1907–1923 (2020)
9. Dunteman, G.H.: Principal components analysis, vol. 69. Sage, New York (1989)
10. Fazio, M., Celesti, A., Ranjan, R., Liu, C., Chen, L., Villari, M.: Open issues in scheduling microservices in the cloud. IEEE Cloud Comput. **3**(5), 81–88 (2016)
11. Filip, I.D., Pop, F., Serbanescu, C., Choi, C.: Microservices scheduling model over heterogeneous cloud-edge environments as support for iot applications. IEEE Internet Things J. **5**(4), 2672–2681 (2018)
12. Foster, I., Zhao, Y., Raicu, I., Lu, S.: Cloud computing and grid computing 360-degree compared. In: 2008 Grid Computing Environments Workshop, pp. 1–10. IEEE, Piscataway (2008)
13. Fu, X., Ghaffar, T., Davis, J.C., Lee, D.: Edgewise: a better stream processing engine for the edge. In: USENIX Annual Technical Conference (ATC) (2019)
14. Han, R., Liu, C.: Cloud-Edge Collaboration with Big Data and Applications. China Machine Press (2022)
15. Han, S., Mao, H., Dally, W.J.: Deep compression: Compressing deep neural networks with pruning, trained quantization and huffman coding (2015). arXiv preprint arXiv:1510.00149
16. Hodge, V., Austin, J.: A survey of outlier detection methodologies. Artif. Intell. Rev. **22**, 85–126 (2004)
17. Insight, T.: Iot connections forecast 2019–2030 (2023). https://transformainsights.com/research/forecast/highlights
18. Jan, B., Farman, H., Khan, M., Imran, M., Islam, I.U., Ahmad, A., Ali, S., Jeon, G.: Deep learning in big data analytics: a comparative study. Comput. Electr. Eng. **75**, 275–287 (2019)
19. Kim, J.O., Mueller, C.W.: Statistical Methods and Practical Issues. Sage Publications, New York (1982)
20. Li, C., Bai, J., Chen, Y., Luo, Y.: Resource and replica management strategy for optimizing financial cost and user experience in edge cloud computing system. Inform. Sci. **516**, 33–55 (2020)
21. Li, E., Zhou, Z., Chen, X.: Edge intelligence: on-demand deep learning model co-inference with device-edge synergy. In: Proceedings of the 2018 Workshop on Mobile Edge Communications, pp. 31–36 (2018)
22. Li, J.: Resource optimization scheduling and allocation for hierarchical distributed cloud service system in smart city. Fut. Gener. Comput. Syst. **107**, 247–256 (2020)
23. Machen, A., Wang, S., Leung, K.K., Ko, B.J., Salonidis, T.: Live service migration in mobile edge clouds. IEEE Wirel. Commun. **25**(1), 140–147 (2017)
24. Ouyang, T., Zhou, Z., Chen, X.: Follow me at the edge: Mobility-aware dynamic service placement for mobile edge computing. IEEE J. Sel. Areas Commun. **36**(10), 2333–2345 (2018)
25. Shi, W., Cao, J., Zhang, Q., Li, Y., Xu, L.: Edge computing: Vision and challenges. IEEE Internet Things J. **3**(5), 637–646 (2016)
26. Shi, W., Zhang, X., Wang, Y., Zhang, Q.: Edge computing: state-of-the-art and future directions. J. Comput. Res. Dev. **56**(1), 69–89 (2019)

27. Strachey, C.: Time sharing in large fast computers. In: Communications of the ACM, vol. 2, pp. 12–13. Association Computing Machinery, New York (1959)
28. Tan, H., Han, Z., Li, X.Y., Lau, F.C.: Online job dispatching and scheduling in edge-clouds. In: IEEE INFOCOM 2017—IEEE Conference on Computer Communications, pp. 1–9. IEEE, Piscataway (2017)
29. Teerapittayanon, S., McDanel, B., Kung, H.T.: Branchynet: Fast inference via early exiting from deep neural networks. In: 2016 23rd International Conference on Pattern Recognition (ICPR), pp. 2464–2469. IEEE, Piscataway (2016)
30. Tong, L., Li, Y., Gao, W.: A hierarchical edge cloud architecture for mobile computing. In: IEEE INFOCOM 2016—The 35th Annual IEEE International Conference on Computer Communications, pp. 1–9. IEEE, Piscataway (2016)
31. Wang, S., Guo, Y., Zhang, N., Yang, P., Zhou, A., Shen, X.: Delay-aware microservice coordination in mobile edge computing: a reinforcement learning approach. IEEE Trans. Mobile Comput. **20**(3), 939–951 (2019)
32. Xiong, Y., Sun, Y., Xing, L., Huang, Y.: Extend cloud to edge with kubeedge. In: 2018 IEEE/ACM Symposium on Edge Computing (SEC), pp. 373–377. IEEE, Piscataway (2018)
33. Yang, H., Zhu, Y., Liu, J.: End-to-end learning of energy-constrained deep neural networks (2018). arXiv e-prints
34. Zhang, J., Ma, M., He, W., Wang, P.: On-demand deployment for iot applications. J. Syst. Archit. **111**, 101794 (2020)
35. Zhang, S., Li, Y., Liu, X., Guo, S., Wang, W., Wang, J., Ding, B., Wu, D.: Towards real-time cooperative deep inference over the cloud and edge end devices. Proc. ACM Interact. Mobile Wearable Ubiquitous Technol. **4**(2), 1–24 (2020)
36. Zhang, T., Li, Z., Chen, Y., Lam, K.Y., Zhao, J.: Edge-cloud cooperation for dnn inference via reinforcement learning and supervised learning. In: 2022 IEEE International Conferences on Internet of Things (iThings) and IEEE Green Computing & Communications (GreenCom) and IEEE Cyber, Physical & Social Computing (CPSCom) and IEEE Smart Data (SmartData) and IEEE Congress on Cybermatics (Cybermatics), pp. 77–84. IEEE, Piscataway (2022)
37. Zou, H., Hastie, T., Tibshirani, R.: Sparse principal component analysis. J. Comput. Graph. Stat. **15**(2), 265–286 (2006)

Chapter 8
Future Prospects

8.1 Brain-Inspired Multimedia Computational Communications

Brain-inspired intelligent communication techniques are critical enablers for next-generation communication networks, aiming to incorporate the human perceptual system and cognitive mechanism into the communication system. To build a comprehensive brain-inspired communication system, the following research challenges should be addressed:

- **Brain-inspired perceptual model of visual quality**: Existing quality evaluation methods for perceptual models measure the similarity of visual features between different objects, producing a score based on an objective assessment system. However, these methods hardly reflect the subjective opinions of users on perceptual quality. To facilitate brain-inspired communications, an electroencephalography (EEG) perceptual model is required to perform human judgment, evaluate visual quality, and guide visual information transmission to improve the quality of subjective user experience. Three bottlenecks need to be addressed for the brain-inspired perceptual model of visual quality: (1) the relationship between visual dimensions and visual perceptual quality: due to the diversity of visual information, the visual quality is generally described in multiple dimensions, e.g., image clarity and video fluency, which prompts to study the relationship between the visual dimensions and the visual quality; (2) EEG perceptual feature extraction: the EEG signals are recorded in extremely short time slots, which increase the difficulty in analyzing the carried information and developing the feature encoder to obtain information related to visual perceptual quality; and (3) a computational model of visual quality: a mathematical theory is necessary to quantitatively map perception dimensions and visual quality degrees to comprehensive user perceptual experience.

X. Tao et al., *Wireless Multimedia Computational Communications*,
Wireless Networks , https://doi.org/10.1007/978-3-031-64155-8_8

- **Brain-inspired cognitive model of visual semantics**: Semantic cognition aims to understand the meaning conveyed by objects. Deep learning (DL)-enabled semantic segment approaches can achieve visual semantic cognition. However, their cognitive performance significantly reduces when visual information is sparse. To develop a brain-inspired visual semantic cognitive system, the following technical problems should be resolved: (1) semantic cognitive limit theory: the constraint relationships among multidimensional EEG signals, signal-to-noise ratio (SNR), and multi-granularity semantics need to be defined to determine the theoretical cognitive boundary of visual semantics; (2) EEG cognitive feature extraction: different from EEG perceptual features, the EEG cognitive features elaborate the characteristics of visual information in the temporal, spatial, and frequency domains, which pose a challenge to the cognitive feature extraction network; and (3) EEG-enabled semantic cognitive system: the relationship between EEG cognitive features and visual semantics is required to build a semantic cognitive model for multimedia data.
- **Computational methodology of brain visual information**: In the conventional transmission paradigm, visual information is compressed, encoded, transmitted over physical channels, and reconstructed at the receiver. In brain-inspired communications, a computational methodology needs to be developed to reveal the coding/decoding process of visual information in the human brain and quantify the visual information before converting it into digital signals for transmission. Three issues need to be addressed: (1) EEG inference feature extraction: the human brain zone that generates visual information should be tracked and identified, and then EEG features corresponding to the essential brain zone are extracted; (2) a mathematical model for EEG feature coding and decoding: a mathematical methodology should be developed to bidirectionally map EEG features to brain visual information based on the extracted EEG features; and (3) EEG-enabled multimedia data reconstruction: a DL-enabled model should be further proposed to reconstruct high-resolution images and videos based on the brain visual information obtained from EEG features.

8.2 Semantic Communications

So far, some initial works on semantic communications have been developed, with most of them focusing on the joint semantic channel coding design to combat semantic errors and channel impairments. To build a comprehensive semantic communication system, the following research challenges should be addressed:

- **Knowledge base**: The first challenge is the evolution of the knowledge base. Knowledge bases consist of large-scale datasets, and their update and maintenance are highly computational and time-consuming. Besides, due to the variation of sources and tasks, intelligent knowledge processing algorithms are required to fuse multimodal semantics inside the knowledge bases, reducing redundant semantic knowledge and correcting corrupted semantic knowledge by achieving autonomous, reactive, and proactive self-evolution. The second challenge is the deployment of the knowledge base. The cloud server deployment structure increases the interaction latency from the knowledge base to the users, while the edge server deployment structure imposes the storage and computation burden on the edge server. The hybrid cloud-edge server deployment structure should be considered, deploying the large-scale shared knowledge base at the cloud server and the unique knowledge base at the edge server to achieve efficient interaction between the knowledge base and users.

- **Affordable semantic-channel coding model**: The DL-enabled semantic-channel coding model is obtained by training complicated neural networks, and processing data with trained models could take up to seconds, making real-time wireless transmission impossible in practice. Many techniques have been utilized to reduce model complexity, such as model compression and network pruning. However, achieving an affordable semantic-channel coding model is still a challenge.
- **Semantic information transmission**: Unlike the conventional Shannon paradigm that transmits bits, semantic communications aim to convey semantic information. New semantic transmission protocols are required to synchronize the semantic representation between the transmitter and the receiver. Additionally, due to the semantic variation of different resources, a general transmission guideline for all resources should be defined, and a robust semantic encoder should be designed to extract semantic information from multimodal data.
- **Comparability with conventional communication systems**: Although DL-enabled semantic communications have shown great potential in breaking technical bottlenecks in conventional communication systems, it is far-fetched to conclude that DL-enabled semantic communications outperform the conventional ones in terms of performance and complexity in any communication scenarios. The joint semantic-channel coding scheme has been demonstrated to be more efficient than the separated scheme, ensuring global optimization and mitigating the overfitting problem of merged neural networks. Moreover, some hybrid semantic-conventional communication architectures have been developed by combining DL-enabled blocks with conventional blocks to yield semantic transmission, but the performance needs to be further verified.

SPRINGER NATURE

GPSR Compliance

The European Union's (EU) General Product Safety Regulation (GPSR) is a set of rules that requires consumer products to be safe and our obligations to ensure this.

If you have any concerns about our products, you can contact us on ProductSafety@springernature.com

In case Publisher is established outside the EU, the EU authorized representative is:

Springer Nature Customer Service Center GmbH
Europaplatz 3
69115 Heidelberg, Germany

The manufacturer's authorised representative in the EU is Springer Nature Customer Service Centre GmbH, Europaplatz 3, 69115 Heidelberg, Germany. If you have any concerns regarding our products, please contact ProductSafety@springernature.com

Printed and bound by CPI Group (UK) Ltd, Croydon, CR0 4YY
01/10/2026
02225950-0005